Fiber Optic and Atmospheric Optical Communication

Fiber Optic and Atmospheric Optical Communication

Nathan Blaunstein
Ben Gurion University of the Negev
Beersheeba, IL

Shlomo Engelberg
Jerusalem College of Technology
Jerusalem, IL

Evgenii Krouk
National Research University Higher School of Economics
Moscow, RS

Mikhail Sergeev
State University of Aerospace Instrumentation
Saint Petersburg, RS

This edition first published 2020
© 2020 John Wiley & Sons, Inc.

The right of Nathan Blaunstein, Shlomo Engelberg, Evgenii Krouk, and Mikhail Sergeev to be identified as the authors of this work has been asserted in accordance with law.

Registered Office
John Wiley & Sons, Inc., 111 River Street, Hoboken, NJ 07030, USA

Editorial Office
111 River Street, Hoboken, NJ 07030, USA

For details of our global editorial offices, customer services, and more information about Wiley products visit us at www.wiley.com.

Wiley also publishes its books in a variety of electronic formats and by print-on-demand. Some content that appears in standard print versions of this book may not be available in other formats.

Library of Congress Cataloging-in-Publication Data applied for

ISBN: 9781119601999

Cover design by Wiley
Cover image: © EasternLightcraft/Getty Images

Set in 10/12pt WarnockPro by SPi Global, Chennai, India

Printed in the United States of America

V10013930_091319

Contents

Preface

This book is intended for scientists, engineers, and designers who would like to learn about optical communications and about the operation and service of optical wireless (atmospheric) and wired (fiber optic) communication links, laser beam systems, and fiber optic multiuser networks. It will be useful to undergraduate and postgraduate students alike and to practicing scientists and engineers.

Over the preceding forty years, many excellent books have been published about different aspects of optical waves and about laser beam propagation in both atmospheric links and within guiding structures, such as fiber optic cables. Wireless and wired communications have often been described separately without taking note of their similarities. In this monograph, we consider both media and describe techniques for transmitting information over such channels when the optical signals are corrupted by the fading that is typical of such communication links because of the existence of artificial (man-made) and/or natural sources of fading.

This monograph methodically unifies the basic concepts and the corresponding mathematical models and approaches to describing optical wave propagation in material media, in waveguide structures and fiber optic structures, and in the troposphere (the lowest layer of the atmosphere). It describes their similarity to other types of electromagnetic waves, e.g. radio waves, from other regions of the electromagnetic spectrum.

Without entering into an overly deep and detailed description of the physical and mathematical fundamentals of the atmosphere as a propagation channel or of the fiber optic structure as a waveguide structure, this monograph focuses the reader's attention on questions related to the coding and decoding that is useful when using such channels. In particular, the monograph analyzes different types of fading and their sources and considers types of modulation that mitigate the effects of fading.

The monograph briefly describes several sources of optical radiation, such as lasers, and presents several particularly relevant optical signal detectors.

The monograph contains material about the atmospheric communication channel, including the effects of atmospheric turbulence and different kinds of hydrometeors, such as aerosols, rain, snow, and clouds, on optical wave propagation in an atmospheric link. The principal goal of this book is to explain the effects of fading and energy loss in information-carrying optical signals. We consider the various situations that occur in the atmospheric link and, finally, show how to mitigate the effects of natural phenomena such as turbulence and hydrometeors that affect the propagation of optical rays and laser beams through the atmosphere.

This book introduces the reader to fading and describes its dispersive nature. It considers fading of optical waves propagating in the irregular turbulent atmosphere in close proximity to the ground surface and elucidates its relation to similar signal dispersive fading phenomena that occur in fiber optic channels where there is a wired link.

The book is organized as follows. Part I consists of two chapters. Chapter 1 describes the fundamental aspects of optical wireless and wired communication links and of the spectrum of optical waves. It also provides a description of the evolution of optical networks (from first to fifth generation networks). End-to-end descriptions of optical channels are provided. Block diagrams of the receiver (the detector of optical waves) and the transmitter (the radiator of optical waves) are given, and information transfer though the channel is described. In Chapter 2, the similarities between radio and optical waves are described. The description makes use of some of the fundamental notions of wave electrodynamics. In particular, the differential and integral presentation of optical waves is developed from Maxwell's equations. Maxwell's equations are also presented in the form of phasors. The principal features of optical wave propagation in material media, both dielectric and conductive, are described. Finally, the reflection and refraction of optical waves from the boundary of two media is described via the introduction of the parameters of refraction (instead of the dielectric and magnetic parameters of the medium), and the effect of total internal reflection, one the main features in any guiding structure (including a fiber optic cable), is considered. There are exercises at the end of Chapter 2.

Part II, which describes the fundamentals of optical communication, consists of six chapters. The first chapter, Chapter 3, describes types of optical signals propagating through wireless or wired communication links. Both continuous and discrete channels are considered, and the relation between them is described. In Chapter 3, we show that for non-correlated optical waves or signals, the average powers of a continuous signal and of a discrete signal (e.g. pulse) are equivalent. The reader is then asked to consider the bandwidth of the signals and to note that one is narrowband and the other wideband. A mathematical/statistical framework is then established for considering these signals in the space, time, and frequency domains. Chapter 4 presents the fundamental principles of discrete signal coding and decoding. The effects of white Gaussian noise on such signals are described briefly and both linear and nonlinear codes are considered. The error probability when decoding such codes is considered for a variety of decoding algorithms for cyclic codes, Reed–Solomon codes, etc. Finally, a general scheme for decoding cyclic codes is developed. In Chapter 5, we apply what we have learned about coding and decoding to optical communication links. Low density parity check codes are considered in detail. Finally, the coding process in optical communication links is described and a comparative analysis of different codes is presented. Chapter 6 considers the effects of fading as it occurs in real optical communication and describes how it is caused by various sources of multiplicative noise. It is shown that by considering signal parameters (pulse duration and bandwidth) and parameters related to channel coherency (in time and frequency), fading phenomena can be described as flat or frequency selective, as slow (in the time domain) or large scale (in the space domain), or as fast (in the time domain) or small scale (in the space domain). Next, mathematical descriptions of fast and slow fading are provided by using the Rayleigh or Rice distribution, gamma-gamma distribution, and the Gaussian distribution. Chapter 7 deals with the modulation of optical signals in wireless and wired communication links. It starts by describing types of analog modulation: analog amplitude modulation and analog phase and frequency modulation, considering them as two types of a general angle modulation of continuous optical signals. The relation between the spectral bandwidths of the two last types of modulation, phase and frequency modulation, is considered, and their signal-to-noise (SNR) ratio is analyzed. Finally, several types of digital signal modulation are presented briefly: amplitude shift keying, phase shift keying, and frequency shift keying. There are several exercises at the end of Chapter 7.

In Chapter 8, optical sources and detectors are described. A brief description of the fundamentals of emission and absorption of optical waves is given. Then the operational characteristics of laser sources and diodes, as well as other types of photodiodes, are briefly described, and several types of modulation schemes that can be used with lasers are demonstrated. Finally, the operational characteristics of photodiodes are presented, and a clear description of the relations between the optical and electrical parameters of typical diode-based schemes is given.

Part III consists of two chapters. In Chapter 9, guiding structures related to fiber optical ones are briefly described. The reader is shown two types of fiber optic structures: step-index fiber and graded-index fiber. Their parameters are determined and described. Next, the propagation of optical waves in fiber optic structures is analyzed, and it is shown that frequency dispersion is an issue when dealing with multimode propagation in such guiding structures. These dispersion properties are examined in Chapter 10, where the corresponding multimode dispersion parameters are presented for the two types of fiber optic cables described in Chapter 9: step-index and graded-index fiber. This modal dispersion is compared with material dispersion caused by the inhomogeneous structure of the material along the length of the fiber. The data loss caused by these two types of dispersion for two kinds of codes – non-return-to-zero (NRZ) codes and return-to-zero (RZ) codes – mentioned in Chapter 1 is described.

Part IV consists of a single chapter. Chapter 11 describes the propagation of optical waves in the atmosphere, considered as an inhomogeneous gaseous structure, and briefly describes the main parameters used to describe the atmosphere. The content of the atmosphere is presented briefly. In particular, in Chapter 11 the structure of aerosols and their dimensions, concentration, spatial distribution of aerosols' sizes, and their spectral extinction and altitude localization are briefly presented. Then, the existence of various water and ice particles, called hydrometeors, in the inhomogeneous atmosphere, their spatial and altitudinal distribution, size distribution, and their effects on optical wave propagation are briefly discussed. Atmospheric turbulent structures caused by temperature and humidity fluctuations combined with turbulent mixing by wind and convection-induced random changes in the air density of the atmosphere (as an irregular gaseous medium) are briefly discussed. Next, the scintillation phenomenon caused by an optical wave passing through the turbulent atmosphere is analyzed. The corresponding formulas for the scintillation index of signal intensity variation are presented as the main parameter of signal fading in the turbulent atmosphere caused by scattering phenomena from turbulent structures. Finally, the corresponding functions used to describe such scattering are described so that the relation between the scintillation index and the fading parameters can be elucidated.

Part V, concerning signal data flow transmission in wireless and fiber optic communication links, consists of one chapter. Chapter 12 starts with definitions related to the characteristics of a communication link: capacity, spectral efficiency, and bit error rate (BER). These important, well-known parameters are presented in a unified manner both for atmospheric and fiber optic channels via the fading parameter, K. Use is made of its relation to the scintillation index that was described in the previous chapter. The relation between the characteristic parameters of the communication link and the fading parameter are described by unified unique formulas and corresponding algorithms. Our understanding of these quantities allows us to perform relevant computations and present clear graphical illustrations for both NRZ and RZ signals.

This book provide a synthesis of several physical and mathematical models in order to present a broad and unified approach for the prediction of data stream parameters for various types of codes used with optical signals traversing optical channels, whether atmospheric or fiber optic, having similar fading time/dispersive effects caused by a variety of sources. In the atmosphere, scattering is due to turbulent structures and hydrometeors; in fiber optic structures, it is due to multimode effects and inhomogeneities in the cladding or core.

Acknowledgments

The authors would like to thank their colleagues for many helpful discussions. They also have the pleasure of acknowledging the computational work of their students – work that led to graphics describing data stream parameters for various situations occurring in wireless atmospheric and wired fiber optic communication channels.

The authors would like to thank the staff at Wiley, the reviewers, and the technical editors for their help in making this book as clear and precise as possible. They would also like to thank Brett Kurzman, Steven Fassioms, and Amudhapriya Sivamurthy of Wiley for their help in bringing this book to market.

The authors are pleased to acknowledge their debt to their families for providing the time, atmosphere, and encouragement that made writing this book such a pleasant undertaking.

Abbreviations

AFD	average fade duration
AM	amplitude modulation
ASK	amplitude shift keying
AWGN	additive white Gaussian noise
BCH	Bose–Chaudhuri–Hocquenghem (codes)
BER	bit error rate
BM	Berlekamp–Massey (iterative algorithm)
BPSK	binary phase shift keying
BSC	binary symmetric channel
BW	bandwidth
CCDF	complementary cumulative distribution function
CDF	cumulative distribution function
C/I	carrier-to-interference ratio
CR	Carson rule
CW	continuous wave
DD	direct detection
EM	electromagnetic (wave, field)
erfc(·)	complementary error function (probability)
F	noise figure
FFF	flat fast fading
FM	frequency modulation
FSF	flat slow fading
FSFF	frequency-selective fast fading
FSK	frequency shift fading
FSSF	frequency-selective slow fading
GIF	gradient-index fiber
HPBW	half power bandwidth
IF	intermediate frequency
Im	imaginary part (of complex number)
IM	intensity modulation
IR	infrared optical spectrum
ISI	inter-symbol interference

LCR	level crossing rate
LD	laser diode
LED	light-emitting diode
LF	likelihood function
LLR	log likelihood ratio
LOS	line-of-sight
LP	linearly polarized (mode)
LPI	low probability of interception
MAD	material dispersion (of fiber optic cable)
MD	multimode dispersion (of fiber optic cable)
MDS	maximum distance separable (codes)
MF	median frequency (band)
NA	numerical aperture of fiber optic guiding structure
NLOS	non-line-of-sight
NRZ	non-return-to-zero (code)
PD	photodiode
PDF	probability density function
PG	processing gain
PGZ	Piterson–Gorenstein–Zincler (algorithm)
PiN	P-type – intrinsic – N-type (detector)
PM	phase modulation
PMD	polarization mode dispersion
PSD	power spectral density
PSK	phase shift keying
QPSK	quadrature phase shift keying
Re	real part (of complex number)
RF	radio frequency (wave or signal)
RS	Reed–Solomon (codes)
RZ	return-to-zero (code)
SIF	step-index fiber
S/N	signal-to-noise ratio
SNR	signal-to-noise ratio
$(SNR)_{in}$	signal-to-noise ratio at the input of the detector
$(SNR)_{out, AM}$	signal-to-noise ratio at the output of the detector for AM signal
$(SNR)_{out, FM}$	signal-to-noise ratio at the output of the detector for FM signal
TE	transverse electric wave
TIR	total intrinsic energy
TM	transverse magnetic wave
UV	ultraviolet optical spectrum
VS	visible optical spectrum
W	energy of depletion zone
W_g	band-gap energy

Nomenclature

$\mathbf{A}$ – arbitrary vectors of electromagnetic field
$\mathbf{B}$ – vector of induction of magnetic field
$\mathbf{E}$ – vector of electric field component of the electromagnetic wave
$\mathbf{E}(z, t)$ – 2-D vector of electrical component of the electromagnetic wave
$\widetilde{\mathbf{E}}(z)$ – phasor of the electrical component of the electromagnetic wave
$\mathbf{D}$ – vector of electric field displacement or vector induction of electric field
$\mathbf{H}$ – vector of magnetic field component of the electromagnetic wave
$\mathbf{H}(z, t)$ – 2-D vector of magnetic field component of the electromagnetic wave
$\widetilde{\mathbf{H}}(z)$ – phasor of the magnetic field component of the electromagnetic wave
$\mathbf{j}$ – vector of electric current density
$\mathbf{J}$ – vector of the full current in medium/circuit
$\mathbf{j}_c$ – conductivity current density
$\mathbf{j}_d$ – displacement current density
$\mathbf{k}$ – wave vector
$\mathbf{M}$ – momentum of the magnetic ambient source
$\mathbf{P}$ – vector of polarization
A_c – amplitude of carrier signal
A_m – amplitude of modulated signal
B_c – coherence bandwidth
B_D – Doppler spread bandwidth
B_f – maximum bandwidth of the modulating signal
B_F – equivalent RF bandwidth of the bandpass filter
B_ω – detector bandwidth
B_Ω – bandwidth of multiplicative noise
$C_{\mathrm{BSC}} = 1 - \eta(p)$ – capacity of binary symmetric channel
p – probability of 0 or 1
f_c – frequency of carrier signal
f_D – Doppler shift
f_m – frequency of modulated signal
B_S – signal bandwidth

C – channel capacity

$C(D)$ – effective cross-section of rain drops as function of their diameter D

$\mathbf{C}_1$ – square $(k \times k)$-matrix

$\mathbf{C}_2$ – matrix of dimension $(k \times n - k)$

$\mathbf{C} = \mathbf{C}_1^{-1}\mathbf{C}_2$ – product of inverse $\mathbf{C}_1$ matrix and regular $\mathbf{C}_2$ matrix

$\mathbf{C}$ – $(m \times m)$-cyclic permutation matrix

$c = \dfrac{1}{\sqrt{\varepsilon_0 \mu_0}}$ – velocity of light in free space

C_n^2 – refraction structure parameter

$C_{\text{NRZ}} \times l = \dfrac{0.7}{\Delta(\tau/l)}$ – capacity per length l of fiber for propagation of NRZ pulses

$C_{\text{RZ}} \times l = 0.875$ (Mbit/c) $\times$ km – capacity per length l of fiber for propagation of RZ pulses

$d\mathbf{l}$ – differential of the vector of a line l

$d\mathbf{S}$ – differential of the vector of a surface S

$d\mathbf{V}$ – differential of the vector of a volume V

$D_m = 0.122 \cdot R^{0.21}$ mm – diameter of rain drops, R – rainfall rate (in mm/h)

D_p – polarization mode dispersion factor (in fiber optic cables)

e – charge of electron

E_b – energy of one transmitted bit

E_x, E_y, E_z – components of the electric field of the wave in the Cartesian coordinate system

$d_H(\mathbf{x}, \mathbf{y})$ – Hamming distance

$\mathbf{G}$ – generator matrix

G – photo-conductive gain of the light detector

$g(t)$ – a signal's envelope as a function of time

I – light intensity

$\mathbf{I}_k$ – unit $(k \times k)$-matrix

$h = 6.625 \cdot 10^{-34}$ J $\cdot$ s – Planck's constant

$h\nu_{ji}$ – energy of photon; j and i are steady states of atoms and electrons, $j > i$

H_W – parity matrix,

$H(\tau) = -\tau \ln \tau - (1 - \tau) \ln(1 - \tau)$ – entropy of the binary ensemble with parameter $\tau = t/n$, $\tau > p$

$i = \sqrt{-1}$ – the unit imaginary number

$J_m(qr)$ – Bessel function of the first kind and of order m

J_{ph} – photocurrent intensity

K – Ricean fading parameter

k_f – frequency deviation constant of frequency modulation

$k_m = (A_m/A_c)$ – modulation index of amplitude modulation

$k = 1.38 \cdot 10^{-23}$ J/K – Boltzmann's constant

$\kappa_0 = \dfrac{2\pi}{L_0}$ – spatial wave number for outer turbulence scale

$\kappa_m = \dfrac{2\pi}{l_0}$ – spatial wave number for inner turbulence scale

k_θ – phase deviation constant of phase modulation

$K_m(qr)$ – modified Hankel function or Bessel function of the second kind and of order m

$K_{\alpha-\beta}[\cdot]$ – modified Bessel function of the second kind of order $(\alpha - \beta)$

L – path loss or attenuation of optical signal

l_0 – inner scale of atmospheric turbulence

L_0 – outer scale of atmospheric turbulence

$l_1 \equiv l_{co} \sim 1/\rho_0$ – coherence length between two coherent points of turbulence

$\ell_F \sim \sqrt{L/k}$ – first Fresnel zone scale, L – range, $k = 2\pi/\lambda$ – wave number

$l_3 \sim R/\rho_0 k$ – scattering disk (turbulence) scale

LP_{01} ($m = 0$) – linear polarized mode with $m = 0$ in fiber optic cable

LP_{11} ($m = 1$) – linear polarized mode with $m = 1$ in fiber optic cable

M – material dispersion factor

$m(x)$ – codeword

$m(t)$ – modulated message signal

$\langle m(t) \rangle$ – average value of the modulated message signal

$M_0 \approx -0.095\,\text{ps}/(\text{nm} \cdot \text{km})$ – material dispersion factor at wavelength of

np – mean optical power

$n(r)$ – aerosol particle distribution in the atmosphere

$n = n' - jn''$ – complex refractive index, $n' = \sqrt{\varepsilon'/\varepsilon_0}$ – real part,

$\quad n'' = \sqrt{\varepsilon''/\varepsilon_0}$ – imaginary part

$n_{\text{eff}} \equiv n_1 \sin \theta_i$ – effective refractive index

$N_0 = 8 \cdot 10^3\,\text{m}^{-2}\,\text{mm}^{-1}$ – constant number of rain drops

N_0 – white noise power spectral density

$N_{\text{add}} = N_0 B_\omega$ – additive (Gaussian) noise power

N_{mult} – spectral density of multiplicative noise

$N_{\text{mult}} = N_{\text{mult}} B_\Omega$ – multiplicative noise power

$N(D)$ – distribution of rain drops as function of their diameter D

$dN(r)$ – number of aerosols having radius between r and $r + dr$

p – pressure in millibars, or pascals or mm Hg

P_r – optical power incident on the detector surface

P_{oIII} – error probability

P_m – mean optical power received by the detector

$P_g(f)$ – PSD of the envelope $g(t)$

P_r – optical signal power

$P_r(e)$ – evaluated probability of the error

$P(h)$ – atmospheric pressure as function of altitude h

$P_m = \langle m^2(t) \rangle$ – power of the modulated message signal

$P(\phi_i) = (2\pi)^{-1}$ – ray phase distribution probability function

R – coefficient of reflection from boundary of two media

R – data rate

R – detector responsivity

$Re = V \cdot l/\nu$ – Reynolds number

r_R – optical ray path length

R_H – coefficient of reflection of the ray with the horizontal polarization

R_S – bulk resistance of the photodetector

R_V – coefficient of reflection of the ray with the vertical polarization

rms $= \sqrt{2} \cdot \sigma \approx 1.414\sigma$ – root mean square

$s(t)$ – bandpass signal

T – temperature in kelvin

T – coefficient of refraction (transfer of the wave into the medium)

$T(h)$ – atmospheric temperature as function of altitude h

t_T – detector transit time

T_b – bit period

T_c – coherence time

$T_p(\tau) = -\tau \ln p - (1-\tau) \ln(1-p)$

T_S – symbol period

$x(t)$ – bandpass signal

$X(f)$ – Fourier transform of $x(t)$

$x_T(t)$ – truncated version of the signal $x(t)$

$X_T(f)$ – Fourier transform of $x_T(t)$

V_p – peak-to-zero value of the modulating signal $m(t)$

v_{ph} – wave phase velocity

$y(t)$ – baseband signal

$Y(f)$ – Fourier transform of $y(t)$

V – rate of data transmission (bits per seconds, bps)

W – energy of arbitrary field

$W_0(f)$ – complex baseband power spectrum in the frequency domain

two media: $n_{\mathrm{eff}} \equiv n_1$ and when $\theta_i = \theta_c$, $n_{\mathrm{eff}} \equiv n_2$

$z = a + ib$ – complex number, a – its real part, b – its imaginary part

$\{x, y, z\}$ – Cartesian coordinate system

$\{\rho, \phi, z\}$ – cylindrical coordinate system

$\{r, \phi, \theta\}$ – spherical coordinate system

∇ – the nabla or del operator for an arbitrary scalar field

$\Delta = \nabla^2$ – Laplacian of the vector or scalar field

Δ – related (fractional) refractive index

$\mathrm{div} = \nabla \cdot$ – divergence of an arbitrary vector (or "del dot the field")

$\mathrm{grad}\, \Phi = \nabla \Phi$ – gradient of arbitrary scalar field or effect of nabla operator on the scalar field

$\mathrm{curl} \equiv \mathrm{rot} = \nabla \times$ – rotor of arbitrary vector field or the vector product of the operator nabla and the field

Δf – peak frequency deviation of the transmitter

α – parameter of wave attenuation in arbitrary medium

$\alpha(\lambda)$ – light scattering coefficient

β – parameter of phase velocity deviation in arbitrary medium

$\beta_f = \Delta f / f_m$ – frequency modulation index

$\beta_\theta = k_\theta A_m = \Delta\theta$ – phase modulation index, k_θ – phase deviation constant of phase modulation

$\gamma = \frac{L}{r} = 4.343\alpha$ – attenuation factor in dB, α – attenuation factor

$\gamma = \alpha + i\beta$ – parameter of propagation in arbitrary material medium

ε – average energy dissipation rate

$\varepsilon = \varepsilon' + i\varepsilon''$ – complex permittivity of arbitrary medium

$\varepsilon_r = \varepsilon_r' + i\varepsilon_r''' = \varepsilon_{Re}' + i\varepsilon_{Im}'''$ – relative permittivity of arbitrary medium, $\varepsilon_{Re}' = \varepsilon'/\varepsilon_0$ – its real part, $\varepsilon_{Im}''' = \varepsilon''/\varepsilon_0$ – its imaginary part

$\varepsilon_0 = \frac{1}{36\pi}10^{-9}$ (F/m) – dielectric parameter of free space

$\Delta\theta$ – peak phase deviation of the transmitter

$\Delta(\tau/l)$ – time delay dispersion of pulses along fiber with the length l

μ_r – mean value of the signal envelope r

μ – permeability of an arbitrary medium

$\mu_0 = 4\pi \cdot 10^{-9}$ (H/m) – permeability of free space

λ – wavelength

$\lambda_0 \approx 1300$ nm – wavelength for material dispersion factor $M = 0$

η – wave impedance in arbitrary medium

$\eta_0 = 120\pi\ \Omega = 377\ \Omega$ – wave impedance of free space

η_B – bandwidth efficiency

η_p – power efficiency

$\eta(p) = -p\log_2 p - (1-p)\log_2(1-p)$ – binary entropy function

$\eta \in N(0, \sigma^2)$ – normally distributed random value with zero mean and variance σ^2

$\delta = \frac{1}{\alpha}$ – skin layer in arbitrary material medium

Φ – photon intensity

λ – wavelength in arbitrary medium

λ_g – wavelength in arbitrary waveguide structure

$K(t, t+\tau) = \langle E(t) \cdot E(t+\tau)\rangle$ – time domain autocorrelation function

ρ – charge density in medium

ρ – density in kg/m^3

$\rho_N(h)$ – density of nitrogen molecules in atmosphere

τ_R – mean electron–hole recombination time

τ_c – correlation time

σ – conductivity of arbitrary medium or material

σ_τ – rms delay spread

$\sigma_r^2(K)$ – variation of the signal envelope r

σ_I^2 – signal intensity scintillation factor

θ_i – angle of incidence of the ray at the boundary of two media

$\theta_c = \sin^{-1}\left(\frac{n_2}{n_1}\right)$ – critical angle of the inner total reflection from boundary of two media

$\theta_{full} = 2 \cdot \theta_a$ – angle of minimum light energy spread outside the cladding of fiber

ϑ – complexity of arrangement of coder/decoder

$\sigma^2 = \langle (\tau - <\tau>)^2 \rangle$ – variance or the mean signal power in the time domain

$\omega = 2\pi f$ – angular frequency

ω_{dn} Doppler shift of the n-ray

Part I

Optical Communication Link Fundamentals

1

Basic Elements of Optical Communication

1.1 Spectrum of Optical Waves

An optical communication system transmits analog and digital information from one place to another using high carrier frequencies lying in the range of 100–1000 THz in the visible and near-infrared (IR) region of the electromagnetic spectrum [1–15]. As for microwave systems, they operate at carrier frequencies that are 5 orders of magnitude smaller ($\sim$1–10 GHz).

White light involves the wavelength band from the ultraviolet spectral band, passing the visible one, to the near-IR band, which has been included here since most fiber communications use carriers in the IR having lowest losses of glass fiber intrinsic surface [8–10].

There are some links in the visible band based on plastic fiber intrinsic surface and therefore having higher loss. Therefore, for the visible optical band such fiber is utilized for short paths. Thus, optical fiber systems operating in 650–670 nm bandwidths with plastic intrinsic surface have loss of 120–160 dB/km, whereas those operating in 800–900 nm bandwidth have loss of 3–5 dB/km, and those operating in 1250–1350 and 1500–1600 nm bandwidth, based on glass surface, have loss of 0.5–0.25 dB/km, respectively.

The relationship between wavelength (λ) and frequency (f) is $\lambda = c/f$, where the velocity of light in free space is $c = 3 \cdot 10^8$ m/s. As an example, a wavelength of $\lambda = 1.5\,\mu$m corresponds to a frequency of $2 \cdot 10^{14}$ Hz $= 2 \cdot 10^2$ THz (a corresponding period of oscillations is $T = 0.5 \cdot 10^{-14}$ s).

A sufficiently wide frequency band has allowed an increase in the bit rate–distance product over a period of about 150 years from $\sim 10^2$ to $\sim 10^{15}$ bps/km (summarized from Refs. [7, 12, 14]).

Fiber Optic and Atmospheric Optical Communication, First Edition.
Nathan Blaunstein, Shlomo Engelberg, Evgenii Krouk, and Mikhail Sergeev.
© 2020 John Wiley & Sons, Inc. Published 2020 by John Wiley & Sons, Inc.

1.2   Optical Communication in Historical Perspective

There are five generations of light communication systems, which differ from each other by wavelength, bit rate (Mbps), and the range of communication between optical terminals [7, 12, 14].

First Generation:

- started in 1980;
- wavelength is 800 nm;
- bit rate – 45 Mbps (megabit per second);
- optical terminals spacing – 10 km.

Second Generation:

- started during the 1980s;
- wavelength is 1300 nm;
- single-mode fiber optic structure;
- bit rate – 1.7 Gbps (gigabit per second);
- optical terminals spacing – 50 km.

Third Generation:

- started in 1990;
- wavelength is 1550 nm;
- single-mode fiber laser;
- dispersive-shifted fibers;
- bit rate – 2.5 Gbps (gigabit per second);
- optical terminals spacing – 60–70 km.

Fourth Generation:

- started in 1996–1997;
- optical amplifiers;
- wavelength-division multiplexing (WDM);
- bit rate – 5–10 Gbps (gigabit per second);
- optical terminals spacing – 60–100 km.

Fifth Generation:

- started at the end of nineties;
- solitons;
- dense-wavelength-division multiplexing (DWDM);
- dispersion compensation;
- optical terminals spacing – up to 100 km.

Figure 1.1 Scheme of optical communication link connected by fiber optics.

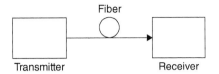

1.3 Optical Communication Link Presentation

Below, we will give a definition of the optical link both for fiber optical link, as a "wire" communication link, and for atmospheric link, a "wireless" communication link. Optical communications via fiber optics can be considered as a finishing optical communication system, as shown in Figure 1.1 according to Refs. [7–12]. The message passing such a link is assumed to be available in electronic form, usually as a current. The transmitter is a light source that is modulated so that the optical beam carries the message.

As an example, for a digital signal, the light beam is electronically turned on (for binary ones) and off (for binary zeros). Here, the optical beam is the carrier of the digital message. Fiber optic links usually take the light-emitting diode and the laser diode as the source. Several characteristics of the light source determine the behavior of the propagating optical waves [1–6]. The corresponding modulated light beam (i.e. the message with the carrier) is coupled into the transmission fiber.

Generally speaking, each optical wireless communication system, wired or wireless, comprises three main blocks: (i) the transmitter, (ii) the channel, and (iii) the receiver (Figure 1.2). The input to the transmitter is an electronic signal, which carries the information, and the output of the transmitter is an optical signal from a light source such as a light-emitting diode (LED) or laser [6, 12] (see also Chapter 8).

The optical signal carries the information presented in digital form as shown in Figure 1.3.

The reader will find full information on the types of digital information presentation in Chapters 3–5. Now we will briefly describe the basic elements of the optical communication channel, including the transmitter, as a source of light, and the receiver, as the detector of light. These elements of the optical communication channel will be discussed in detail in Chapter 8.

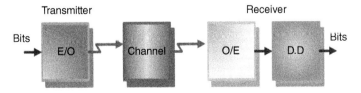

Figure 1.2 Optical wired or wireless system scheme.

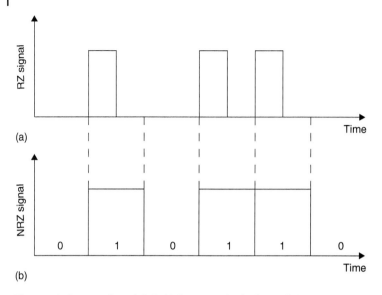

Figure 1.3 Presentation of digital information in the form of (a) return-to-zero (RZ) and (b) non-return-to-zero (NRZ) pulses.

The input to the channel is the optical signal from the transmitter and the output of the channel is the input to the receiver. The receiver receives the optical signal from the output of the channel, amplifies the signal, converts it to an electronic signal, and extracts the information. At the receiver, the signal is collected by a photodetector, which converts the information back into electrical form. The photodetectors do not affect the propagation properties of the optical wave but certainly must be compatible with the rest of the system (see Chapter 8).

The transmitter includes a modulator, a driver, a light source, and optics (Figure 1.4). The modulator converts the information bits to an analog signal that represents a symbol stream. The driver provides the required current to the light source based on the analog signal from the output of the modulator. The light source is an LED or a laser, which is a noncoherent or a coherent source, respectively. The source converts the electronic signal to an optical signal [6, 12]. The optics focuses and directs the light from the output of the source in the direction of the receiver.

The receiver includes optics, a filter, a polarizer, a detector, a trans-impedance amplifier, a clock recovery unit, and a decision device (see Figure 1.5). The optics concentrates the received signal power onto the filter.

Only light at the required wavelength propagates through the filter to the polarizer. The polarizer only enables light at the required polarization to propagate through to the detector. The detector, in most cases a semiconductor

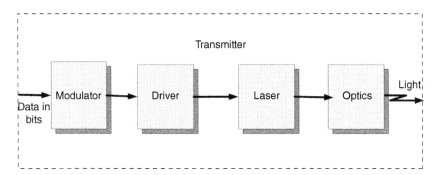

Figure 1.4 The transmitter scheme.

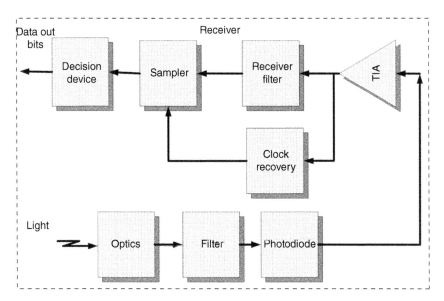

Figure 1.5 The receiver scheme.

device such as a positive intrinsic negative (PIN) photodiode, converts the optical signal to an electronic signal (see Chapter 8). The trans-impedance amplifier amplifies the electronic signal from the detector. The clock recovery unit provides a synchronization signal to the decision device based on the signal from the output of the trans-impedance amplifier. The decision device estimates the received information based on the electronic signal from the trans-impedance amplifier and synchronization signal.

The atmospheric (wireless) channel attenuates the power of the optical signal and widens and spreads it in the spatial, temporal, angular, and polarization domains (see Figure 1.6) [13, 15].

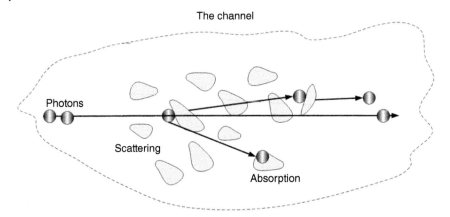

Figure 1.6 The atmospheric optical channel.

The attenuation, widening, and spreading are stochastic processes, resulting from the interaction of light with atmospheric gases, aerosols, and turbulence. The atmospheric gases mostly absorb the light while the aerosols absorb and scatter the light. The turbulence gives rise to constructive and destructive interference, which cause fluctuations in received power, or scintillations, as well as beam bending in the detector plane, as described in Chapters 10 and 11.

References

1 Jenkis, F.A. and White, H.E. (1953). *Fundamentals of Optics*. New York: McGraw-Hill.
2 Born, M. and Wolf, E. (1964). *Principles in Optics*. New York: Pergamon Press.
3 Fain, V.N. and Hanin, Y.N. (1965). *Quantum Radiophysics*. Moscow: Sov. Radio (in Russian).
4 Akhamov, S.A. and Khohlov, R.V. (1965). *Problems of Nonlinear Optics*. Moscow: Fizmatgiz (in Russian).
5 Lipson, S.G. and Lipson, H. (1969). *Optical Physics*. Cambridge: University Press.
6 Akhamov, S.A., Khohlov, R.V., and Sukhorukov, A.P. (1972). *Laser Handbook*. North Holland: Elsevier.
7 Marcuse, O. (1972). *Light Transmission Optics*. New York: Van Nostrand-Reinhold Publisher.
8 Kapany, N.S. and Burke, J.J. (1972). Optical Waveguides, Chapter 3. In: New York: Academic Press.
9 Fowles, G.R. (1975). *Introduction in Modern Optics*. New York: Holt, Rinehart, and Winston Publishers.

10 Midwinter, J.E. (1979). *Optical Fibers for Transmission*. New York: Wiley.

11 Hecht, E. (1987). *Optics*. Boston, MA: Addison-Wesley, Reading.

12 Dakin, J. and Culshaw, B. (eds.) (1988). *Optical Fiber Sensors: Principles and Components*. Boston-London: Artech House.

13 Kopeika, N.S. (1998). *A System Engineering Approach to Imaging*. Washington: SPIE Optical Engineering Press.

14 Bansal, R. (ed.) (2006). *Handbook: Engineering Electromagnetics Applications*. New York: Taylor & Francis.

15 Blaunstein, N., Arnon, S., Zilberman, A., and Kopeika, N. (2010). *Applied Aspects of Optical Communication and LIDAR*. New York: CRC Press, Taylor & Francis Group.

2

Optical Wave Propagation

2.1 Similarity of Optical and Radio Waves

Electromagnetic (EM) wave plays a key role in modern communication systems, radio and optical, including optical telecommunication systems [1–6]. Therefore, fiber optic or laser communications emphasize the EM phenomena described mathematically by Maxwell's unified theory [7–17], which we will consider in detail subsequently. Investigation of optical communications started just after the invention of the laser in 1960. Atmospheric propagation of optical waves was investigated parallelly with improvements in optical lasers and understanding of the problems of optical wave's generation both in the time and space domains. Thus, problems with weather, line-of-site (LOS) clearance, beam spreading and bending in the real atmosphere consisting of gaseous structures (aerosols, fog, smoke, turbulence, etc.), and hydrometeors (snow, rain, clouds, etc.) eliminated free-space optical communication as a major aspect in wireless communication.

In the middle of 1960s, guided propagation in a glass fiber was proposed as a strategy for overcoming the many problems of optical atmospheric communications, which led to understanding of all processes, such as multi-scattering, multi-reflection, and multi-diffraction, as well as refraction and fading effects caused by inhomogeneous structures existing in the real irregular atmosphere. In the seventies, the first highly transparent glass fiber was produced, making fiber optic communication practical.

Since optical waves have the same nature as EM waves, having their own part in the frequency (or wavelength) domain (see Chapter 1), we will start with a physical explanation of the EM wave based on Maxwell's unified theory [7, 8, 10–13], which postulates that an EM field could be represented as a wave. The coupled wave components, the electric and magnetic fields, are depicted in Figure 2.1, from which it follows that the EM wave travels in a direction perpendicular to both EM field components. In Figure 2.1 this direction is denoted as the z-axis in the Cartesian coordinate system by the wave vector **k**. In their orthogonal space planes, the magnetic and electric oscillatory

Fiber Optic and Atmospheric Optical Communication, First Edition.
Nathan Blaunstein, Shlomo Engelberg, Evgenii Krouk, and Mikhail Sergeev.
© 2020 John Wiley & Sons, Inc. Published 2020 by John Wiley & Sons, Inc.

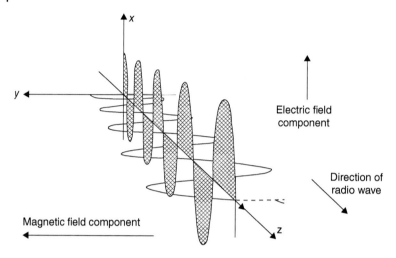

Figure 2.1 Optical wave as an electromagnetic wave with its electrical and magnetic components, wavefront, and direction of propagation presentation.

components repeat their waveform after a distance of one wavelength along the y-axis and x-axis, respectively (see Figure 2.1).

Both components of the EM wave are in phase in the time domain, but not in the space domain [7, 8, 10–13]. Moreover, the magnetic component value of the EM field is closely related to the electric component value, from which one can obtain the radiated power of the EM wave propagating along the z-axis (see Figure 2.1).

At the same time, using the Huygens principle, well known in electrodynamics [10–13], one can show that optical wave is the EM wave propagating only straight from the source as rays with minimum loss of energy and with minimum time spent for propagation (according to Principe of the Fermat [4, 8, 15]) in free space, as an unbounded homogeneous medium without obstacles and discontinuities.

Thus, if we present the Huygens concept, as shown in Figure 2.2, the ray from each point propagates in all forward directions to form many elementary spherical wavefronts, which Huygens called wavelets.

The envelope of these wavelets forms the new wavefronts. In other words, each point on the wavefront acts as a source of secondary elementary spherical waves, described by Green's function (see Refs. [10–13]). These waves combine to produce a new wavefront in the direction of wave propagation in a straight manner. As will be shown below, each wavefront can be represented by the plane that is normal to the wave vector **k** (e.g. wave energy transfer). Moreover, propagating forward along straight lines normal to their wavefront the waves propagate as light rays in optics, spending minimum energy for passing from the source to any detector, that is, the maximum energy of the ray is observed

Figure 2.2 Huygens principle for a proof of straight propagation of waves as rays.

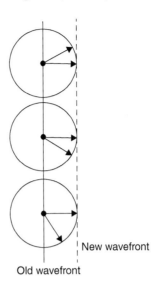

New wavefront

Old wavefront

in the straight direction normal to the wavefront (as seen from Figure 2.2). Kirchhoff was the first to mathematically show this principle, based on the general Maxwell's unified theory. Let us now assess all propagation phenomena theoretically using Maxwell's unified theory.

2.2 Electromagnetic Aspects of Optical Wave Propagation

The theoretical analysis of optical wave propagation, as a part of the whole EM spectrum [1–3, 7–9] (see also Chapter 1), is based on Maxwell's equations [10–17]. In vector notation and in the SI units system, their representations in the uniform macroscopic form are [1–3, 7–9]

$$\nabla \times \mathbf{E}(\mathbf{r}, t) = -\frac{\partial}{\partial t}\mathbf{B}(\mathbf{r}, t) \tag{2.1a}$$

$$\nabla \times \mathbf{H}(\mathbf{r}, t) = \frac{\partial}{\partial t}\mathbf{D}(\mathbf{r}, t) + \mathbf{j}(\mathbf{r}, t) \tag{2.1b}$$

$$\nabla \cdot \mathbf{B}(\mathbf{r}, t) = 0 \tag{2.1c}$$

$$\nabla \cdot \mathbf{D}(\mathbf{r}, t) = \rho(\mathbf{r}, t) \tag{2.1d}$$

Here, $\mathbf{E}(\mathbf{r}, t)$ is the electric field strength vector, in volts per meter (V/m); $\mathbf{H}(\mathbf{r}, t)$ is the magnetic field strength vector, in amperes per meter (A/m); $\mathbf{D}(\mathbf{r}, t)$ is the electric flux induced in the medium by the electric field, in coulombs per cubic meter (C/m^3) (this is why in the literature sometimes it is called an "induction" of an electric field); $\mathbf{B}(\mathbf{r}, t)$ is the magnetic flux induced by the magnetic field, in webers per square meter (Wb/m^2) (it is also called "induction" of a magnetic field); $\mathbf{j}(\mathbf{r}, t)$ is the vector of electric current density, in amperes per square meter (A/m^2); $\rho(\mathbf{r}, t)$ is the charge density in coulombs per square meter (C/m^2). The curl operator, $\nabla\times$, is a measure of field rotation, and the divergence operator, $\nabla\cdot$, is a measure of the total flux radiated from a point.

It should be noted that for a time-varying EM-wave field, Eqs. (2.1c) and (2.1d) can be derived from (2.1a) and (2.1b), respectively. In fact, taking the divergence of (2.1a) (by use of the divergence operator $\nabla\cdot$) one can immediately obtain (2.1c). Similarly, taking the divergence of (2.1b) and using the well-known continuity equation [7–13]

$$\nabla \cdot \mathbf{j}(\mathbf{r}, t) + \frac{\partial \rho(\mathbf{r}, t)}{\partial t} = 0 \tag{2.2}$$

one can arrive at (2.1d). Hence, only two equations, Eqs. (2.1a) and (2.1b), are independent.

Equation (2.1a) is the well-known Faraday law and indicates that a time-varying magnetic flux generates an electric field with rotation; (2.1b) without the term $\frac{\partial \mathbf{D}}{\partial t}$ (displacement current term [10–13]) limits to the well-known Ampere law and indicates that a current- or a time-varying electric flux (displacement current [10–13]) generates a magnetic field with rotation.

Because one now has only two independent equations, Eqs. (2.1a) and (2.1b), which describe the four unknown vectors $\mathbf{E}$, $\mathbf{D}$, $\mathbf{H}$, $\mathbf{B}$, three more equations relating these vectors are needed. To do this, we introduce relations between $\mathbf{E}$ and $\mathbf{D}$, $\mathbf{H}$ and $\mathbf{B}$, $\mathbf{j}$ and $\mathbf{E}$, which are known in electrodynamics. In fact, for isotropic media, which are usually considered in problems of land radio propagation, the electric and magnetic fluxes are related to the electric and magnetic fields, and the electric current is related to the electric field via the constitutive relations [10–13]:

$$\mathbf{B} = \mu(\mathbf{r})\mathbf{H} \tag{2.3}$$

$$\mathbf{D} = \varepsilon(\mathbf{r})\mathbf{E} \tag{2.4}$$

$$\mathbf{j} = \sigma(\mathbf{r})\mathbf{E} \tag{2.5}$$

It is very important to note that relations (2.4)–(2.5) are valid only for propagation processes in linear isotropic media, which are characterized by the three scalar functions of any point $\mathbf{r}$ in the medium: permittivity $\varepsilon(\mathbf{r})$, permeability $\mu(\mathbf{r})$ and conductivity $\sigma(\mathbf{r})$. In that relations (2.3)–(2.5) it was assumed that the medium is inhomogeneous. In a homogeneous medium the functions $\varepsilon(\mathbf{r})$, $\mu(\mathbf{r})$, and $\sigma(\mathbf{r})$ transform to simple scalar values ε, μ, and σ.

In free space, these functions are simply constants, i.e. $\varepsilon = \varepsilon_0 = 8.854 \cdot 10^{-12} = \frac{1}{36\pi} 10^{-9}$ F/m, while $\mu = \mu_0 = 4\pi \cdot 10^{-7}$ H/m. The constant $c = \frac{1}{\sqrt{\varepsilon_0 \mu_0}} = 3 \times 10^8$ m/s is the velocity of light, which has been measured very accurately.

The system (2.5) can be further simplified if we assume that the fields are time harmonic. If the field time dependence is not harmonic, then, using the fact that Eqs. (2.1) are linear, we may treat these fields as sums of harmonic components and consider each component separately. In this case, the time harmonic field is a complex vector and can be expressed via its real part as [10–13, 17]

$$\mathbf{A}(\mathbf{r}, t) = Re\left[\mathbf{A}(\mathbf{r})e^{-i\omega t}\right] \tag{2.6}$$

where $i = \sqrt{-1}$, ω is the angular frequency in radians per second, $\omega = 2\pi f$, f is the radiated frequency (in Hz $= s^{-1}$), and $\mathbf{A}(\mathbf{r}, t)$ is the complex vector ($\mathbf{E}$, $\mathbf{D}$, $\mathbf{H}$, $\mathbf{B}$, or $\mathbf{j}$). The time dependence $\sim e^{-i\omega t}$ is commonly used in the literature of electrodynamics and wave propagation. If $\sim e^{i\omega t}$ is used, then one must substitute $-i$ for i and i for $-i$, in all equivalent formulations of Maxwell's equations. In (2.6) $e^{-i\omega t}$ presents the harmonic time dependence of any complex vector $\mathbf{A}(\mathbf{r}, t)$, which satisfies the relationship

$$\frac{\partial}{\partial t} \mathbf{A}(\mathbf{r}, t) = Re\left[-i\omega \mathbf{A}(\mathbf{r})e^{-i\omega t}\right] \tag{2.7}$$

Using this transformation, one can easily obtain from the system (2.1)

$$\nabla \times \mathbf{E}(\mathbf{r}) = i\omega \mathbf{B}(\mathbf{r}) \tag{2.8a}$$

$$\nabla \times \mathbf{H}(\mathbf{r}) = -i\omega \mathbf{D}(\mathbf{r}) + \mathbf{j}(\mathbf{r}) \tag{2.8b}$$

$$\nabla \cdot \mathbf{B}(\mathbf{r}) = 0 \tag{2.8c}$$

$$\nabla \cdot \mathbf{D}(\mathbf{r}) = \rho(\mathbf{r}) \tag{2.8d}$$

It can be observed that system (2.7) was obtained from system (2.1) by replacing $\partial/\partial t$ with $-i\omega$. Alternatively, the same transformation can be obtained by the use of the Fourier transform of system (2.1) with respect to time [7, 8, 10–15]. In Eqs. (2.8a)–(2.8d) all vectors and functions are actually the Fourier transforms with respect to the *time domain*, and the fields $\mathbf{E}$, $\mathbf{D}$, $\mathbf{H}$, and $\mathbf{B}$ are functions of frequency as well; we call them *phasors* of time domain vector solutions. They are also known as the *frequency domain solutions* of the EM field according to system (2.8). Conversely, the solutions of system (2.1) are the *time domain solutions* of the EM field. It is more convenient to work with system (2.8) instead of system (2.1) because of the absence of the time dependence and time derivatives in it.

2.3   Propagation of Optical Waves in Free Space

Mathematically, optical wave propagation phenomena can be described by the use of both the scalar and vector wave equation presentations. Because most problems of optical wave propagation in wireless communication links are considered in unbounded, homogeneous, source-free isotropic media, we can consider $\varepsilon(\mathbf{r}) \equiv \varepsilon$, $\mu(\mathbf{r}) \equiv \mu$, $\sigma(\mathbf{r}) \equiv \sigma$, and finally obtain from the general wave equations,

$$\nabla \times \nabla \times \mathbf{E}(\mathbf{r}) - \omega^2 \varepsilon \mu \mathbf{E}(\mathbf{r}) = 0$$

$$\nabla \times \nabla \times \mathbf{H}(\mathbf{r}) - \omega^2 \varepsilon \mu \mathbf{H}(\mathbf{r}) = 0 \qquad (2.9)$$

Because both equations are symmetric, one can use one of them, namely, that for **E**, and by introducing the vector relation $\nabla \times \nabla \times \mathbf{E} = \nabla(\nabla \cdot \mathbf{E}) - \nabla^2 \mathbf{E}$ and taking into account that $\nabla \cdot \mathbf{E} = 0$, finally obtain

$$\nabla^2 \mathbf{E}(\mathbf{r}) + k^2 \mathbf{E}(\mathbf{r}) = 0 \qquad (2.10)$$

where $k^2 = \omega^2 \varepsilon \mu$.

In special cases of a homogeneous, source-free, isotropic medium, the three-dimensional wave equation reduces to a set of scalar wave equations. This is because in Cartesian coordinates, $\mathbf{E}(\mathbf{r}) = E_x \mathbf{x}_0 + E_y \mathbf{y}_0 + E_z \mathbf{z}_0$, where $\mathbf{x}_0$, $\mathbf{y}_0$, $\mathbf{z}_0$ are unit vectors in the directions of the x, y, z coordinates, respectively. Hence, Eq. (2.10) consists of three scalar equations such as

$$\nabla^2 \Psi(\mathbf{r}) + k^2 \Psi(\mathbf{r}) = 0 \qquad (2.11)$$

where $\psi(\mathbf{r})$ can be E_x, E_y, or E_z. This equation fully describes propagation of optical wave in free space.

2.4   Propagation of Optical Waves Through the Boundary of Two Media

2.4.1   Boundary Conditions

The simplest case of wave propagation over the intersection between two media is that where the intersection surface can be assumed to be flat and perfectly conductive.

If so, for a perfectly conductive flat surface the total electric field vector is equal to zero, i.e. $\mathbf{E} = 0$ [7–13, 17]. In this case, the tangential component of the electric field vanishes at the perfectly conductive flat surface, that is,

$$E_\tau = 0 \qquad (2.12)$$

Consequently, as follows from Maxwell's equation $\nabla \times \mathbf{E}(\mathbf{r}) = i\omega \mathbf{H}(\mathbf{r})$ (see Eq. (2.8a) for the case of $\mu = 1$ and $\mathbf{B} = \mathbf{H}$), at such a flat perfectly conductive material surface the normal component of the magnetic field also vanishes, i.e.

$$H_n = 0 \qquad (2.13)$$

As also follows from Maxwell's equations (2.1), the tangential component of magnetic field does not vanish because of its compensation by the surface electric current. At the same time, the normal component of the electric field is also compensated by pulsing electrical charge at the material surface. Hence, by introducing the Cartesian coordinate system, one can present the boundary conditions (2.12)–(2.13) at the flat perfectly conductive material surface as follows:

$$E_x(x, y, z = 0) = E_y(x, y, z = 0) = H_z(x, y, z = 0) = 0 \qquad (2.14)$$

2.4.2 Main Formulations of Reflection and Refraction Coefficients

As was shown above, the influence of a flat material surface on optical wave propagation leads to phenomena such as reflection. Because all kinds of waves can be represented by means of the concept of plane waves [7–13], let us obtain the main reflection and refraction formulas for a plane wave incident on a plane surface between two media, as shown in Figure 2.3. The media have different dielectric properties, which are described above and below the boundary plane $z = 0$ by the permittivity and permeability ε_1, μ_1 and ε_2, μ_2, respectively for each medium.

Without reducing the general problem, let us consider an optical wave with wave vector $\mathbf{k}_0$ and frequency $\omega = 2\pi f$ incident from a medium described by parameter n_1. The reflected and refracted waves are described by wave vectors $\mathbf{k}_1$ and $\mathbf{k}_2$, respectively. Vector $\mathbf{n}$ is a unit normal vector directed from a medium with n_2 into a medium with n_1.

We should notice that in optics usually designers of optical systems deal with non-magnetized materials assuming the normalized dimensionless

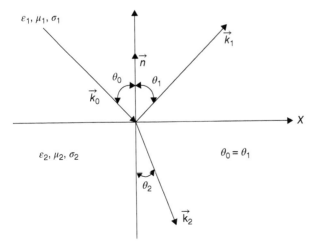

Figure 2.3 Reflection and refraction of optical wave at the boundary of two media.

permeability of the two media equal to unity, that is, $\mu_1 = \tilde{\mu}_1/\mu_0 = 1$ and $\mu_2 = \tilde{\mu}_2/\mu_0 = 1$. Moreover, instead of the normalized dimensionless permittivity for each medium, $\varepsilon_1 = \tilde{\varepsilon}_1/\varepsilon_0$, $\varepsilon_2 = \tilde{\varepsilon}_2/\varepsilon_0$, that is usually used in electrodynamics and radio physics, they are expressed via the corresponding refractive index n_1 and n_2, that is, $\varepsilon_1 = n_1^2$ and $\varepsilon_2 = n_2^2$. According to the relations between electrical and magnetic components, which follow from Maxwell's equations (see system (2.1)), we can easily obtain the expressions for the coefficients of reflection and refraction. A physical meaning of the reflection coefficient is as follows: it defines the ratio of the reflected electric field component of the optical wave to its incident electric filed component. The same physical meaning is applicable to the refractive coefficient: it defines the ratio of the refractive electric field component to the incident electric field component of the optical wave. Before presenting these formulas, let us show two important laws usually used in classical optics. As follows from Maxwell's laws, the boundary conditions, and the geometry presented in Figure 2.3, the values of the wave vectors are related by the following expressions [17]:

$$| \mathbf{k}_0 |=| \mathbf{k}_1 | \equiv k = \frac{\omega}{c} n_1, \quad | \mathbf{k}_2 | \equiv k_2 = \frac{\omega}{c} n_2 \tag{2.15}$$

From the boundary conditions that were described earlier by (2.12)–(2.14), one can easily obtain the condition of the equality of phase for each wave at the plane $z = 0$:

$$(\mathbf{k}_0 \cdot \mathbf{x})_{z=0} = (\mathbf{k}_1 \cdot \mathbf{x})_{z=0} = (\mathbf{k}_2 \cdot \mathbf{x})_{z=0} \tag{2.16}$$

which is independent of the nature of the boundary condition. Equation (2.16) describes the condition that all three wave vectors must lie in the same plane. From this equation it also follows that

$$\mathbf{k}_0 \sin \theta_0 = \mathbf{k}_1 \sin \theta_1 = \mathbf{k}_2 \sin \theta_2 \tag{2.17}$$

which is the analog of the *second Snell's law*:

$$n_1 \sin \theta_0 = n_2 \sin \theta_2 \tag{2.18}$$

Moreover, because $|\mathbf{k}_0| = |\mathbf{k}_1|$, we find $\theta_0 = \theta_1$; the angle of incidence equals the angle of reflection. This is the *first Snell's law*.

In the literature that describes wave propagation aspects, optical waves are usually called the waves with *vertical* and *horizontal* polarization, depending on the orientation of the electric field component with respect to the plane of propagation, perpendicular or parallel, respectively.

Without entering into straight retinue computations, following classical electrodynamics, we will obtain the expressions for the complex coefficients of reflection (R) and refraction (T) for waves with vertical (denoted by index V) and horizontal (denoted by index H) polarization, respectively.

For *vertical* polarization:

$$R_V = \frac{n_1 \cos\theta_0 - \sqrt{n_2^2 - n_1^2\sin^2\theta_0}}{n_1 \cos\theta_0 + \sqrt{n_2^2 - n_1^2\sin^2\theta_0}} \tag{2.19a}$$

$$T_V = \frac{2n_1 \cos\theta_0}{n_1 \cos\theta_0 + \sqrt{n_2^2 - n_1^2\sin^2\theta_0}} \tag{2.19b}$$

For *horizontal* polarization:

$$R_H = \frac{-n_2^2 \cos\theta_0 + n_1\sqrt{n_2^2 - n_1^2\sin^2\theta_0}}{n_2^2 \cos\theta_0 + n_1\sqrt{n_2^2 - n_1^2\sin^2\theta_0}} \tag{2.20a}$$

$$T_H = \frac{2n_1 n_2 \cos\theta_0}{n_2^2 \cos\theta_0 + n_1\sqrt{n_2^2 - n_1^2\sin^2\theta_0}} \tag{2.20b}$$

Dependence of the coefficient of reflection on the angle of incidence is shown in Figure 2.4, according to [15], for two types of field polarization, R_V and R_H.

In the case of vertical polarization there is a special angle of incidence, called the *Brewster angle*, for which there is no reflected wave. For simplicity, we will assume that the condition $\mu_1 = \mu_2$ is valid. Then, from (2.18) it follows that the reflected wave limits to zero when the angle of incidence is equal to Brewster's angle

$$\theta_0 \equiv \theta_{Br} = \tan^{-1}\left(\frac{n_2}{n_1}\right) \tag{2.21}$$

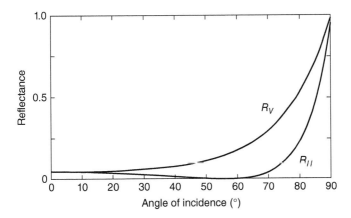

Figure 2.4 Reflectance for both types of polarization vs. the angle of incidence.

Another interesting phenomenon that follows from the presented formulas is called *total reflection*. It takes place when the condition $n_2 \gg n_1$ is valid. In this case, from Snell's law (2.18) it follows that, if $n_2 \gg n_1$, then $\theta_1 \gg \theta_i$. Consequently, when $\theta_i \gg \theta_c$ the reflection angle $\theta_1 = \frac{\pi}{2}$, where

$$\theta_c = \sin^{-1}\left(\frac{n_2}{n_1}\right) \tag{2.22}$$

For waves incident at the surface under the critical angle $\theta_i \equiv \theta_c$, there is no refracted wave within the second medium; the refracted wave is propagated along the boundary between the first and second media and there is no energy flow across the boundary of these two media.

Therefore, this phenomenon is called *total internal reflection* (TIR) in the literature, and the smallest incident angle θ_i for which we get TIR is called the critical angle $\theta_i \equiv \theta_c$ defined by expression (2.22).

2.5 Total Intrinsic Reflection in Optics

We can rewrite now a Snell's, presented above for $\theta_i = \theta_1$, as [1–3, 7–9] (see also the geometry of the problem shown in Figure 2.3)

$$n_1 \sin \theta_r = n_2 \sin \theta_t \tag{2.23}$$

or

$$\sin \theta_i \equiv \sin \theta_r = \frac{n_2}{n_1} \sin \theta_t \tag{2.24}$$

If the second medium is less optically dense than the first medium, which consists of the incident ray with amplitude $|E_i|$, that is, $n_1 > n_2$, from (2.24) it follows that

$$\sin \theta_i > \frac{n_2}{n_1} \tag{2.25a}$$

or

$$\frac{n_1}{n_2} \sin \theta_i > 1 \tag{2.25b}$$

The value of the incident angle θ_i for which (2.25) becomes true is known as a *critical angle*, which was introduced above. We now define its meaning by the use of a ray concept [7–9]. If a critical angle is determined by (2.22), which we will rewrite in another manner,

$$\sin \theta_c = \frac{n_2}{n_1} \tag{2.26}$$

Then, for all values of incident angles $\theta_i > \theta_c$ the light is totally reflected at the boundary of two media. This phenomenon is called in ray theory the TIR

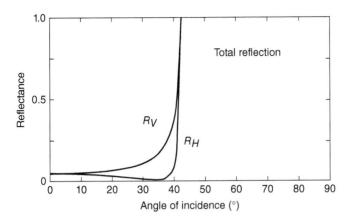

Figure 2.5 Total reflection at the angle of incidence $\theta_i > \theta_c \approx 41° - 42°$.

of rays, the effect that is very important in light propagation in fiber optics. Figure 2.5, following [15], shows the effect of total reflection when $\theta_i > \theta_c$.

We also can introduce another main parameter usually used in optic communication. Thus, the *effective index of refraction* is defined as $n_{\text{eff}} \equiv n_1 \sin \theta_i$. When the incident ray angle $\theta_i = 90°$, $n_{\text{eff}} \equiv n_1$, and when $\theta_i = \theta_c$, $n_{\text{eff}} \equiv n_2$.

Let us now explain the TIR from another point of view. When we have TIR, we should assume that there would be no electric field in the second medium. This is not the case, however. The boundary conditions presented above require that the electric field be continuous at the boundary; that is, at the boundary the field in region 1 and that in region 2 must be equal. The exact solution shows that due to TIR we have in region 1 standing waves caused by interference of incident and fully reflected waves, whereas in region 2 a finite electric field decays exponentially away from boundary and carries no power into the second medium. This wave is called an *evanescent field* (see Figure 2.4).

This field attenuates away from the boundary as

$$E \propto \exp\{-\alpha z\} \tag{2.27}$$

where the attenuation factor equals

$$\alpha = \frac{2\pi}{\lambda}\sqrt{n_1^2\sin^2\theta_i - n_2^2} \tag{2.28}$$

As can be seen from (2.28), at the critical angle $\theta_i = \theta_c$, $\alpha \to 0$, and attenuation increases as the incident angle increases beyond the critical angle defined by (2.26). Because α is so small near the critical angle, the evanescent fields penetrate deeply beyond the boundary but do so less and less as the angle increases.

However, the behavior of the main formulas (2.19)–(2.20) depends on boundary conditions. Thus, if the fields are to be continuous across the boundary, as required by Maxwell's equations, there must be a field disturbance of some kind in the second media (see Figure 2.2). To investigate this disturbance we can use Fresnel's formulas. We first rewrite, following [15], $\cos\theta_t = (1 - \sin^2\theta_t)^{1/2}$. For $\theta_t > \theta_c$, we can present $\sin\theta_t$ by the use of some additional function $\sin\theta_t = \cosh\gamma$, which can be more than unity. If so, $\cos\theta_t = j(\cosh^2\gamma - 1)^{1/2} = \pm j\sinh\gamma$. Hence, we can write the field component in the second medium to vary as (for nonmagnetic materials $\mu_1 = \mu_2 = \mu_0$)

$$\exp\left\{j\omega\left(t - n_2\frac{x\cosh\gamma - jz\sinh\gamma}{c}\right)\right\} \tag{2.29a}$$

or

$$\exp\left(-\frac{\omega n_2 z\sinh\gamma}{c}\right)\exp\left\{j\omega\left(t - n_2\frac{x\cosh\gamma}{c}\right)\right\} \tag{2.29b}$$

This formula represents a ray traveling in the z-direction in the second medium (that is, parallel to the boundary) with amplitude decreasing exponentially in the z-direction (at right angles to the boundary). The rate of the amplitude decrease in the z-direction can be written as

$$\exp\left(-n_2\frac{2\pi z\sinh\gamma}{\lambda_2}\right)$$

where λ_2 is the wavelength of light in the second medium. The wave with exponential decay is usually called *evanescent wave* in the literature [15, 16]. As seen from Figure 2.6, rearranged from [16], the wave attenuates significantly ($\sim e^{-1}$) over a critical distance d_c of about λ_2.

Another expression of the evanescent wave decay region, d_c, can be obtained, following Ref. [15], by introducing the incident angle of light at the boundary of two media θ and both refractive indexes of the media, n_1 and n_2:

$$d_c = \frac{\lambda_2}{2\pi(n_1^2\sin^2\theta - n_2^2)^{1/2}} \tag{2.30}$$

This critical depth of field exponential attenuation is similar to the characteristics of EM wave penetration into the material usually used in electrodynamics and electromagnetism, and called the *skin layer* [11–13].

As for the left side of Figure 2.6, the standing wave occurs as a result of interaction between two optical waves, the incident and the reflected waves from the interface of two media. We should notice that this picture is correct in situations

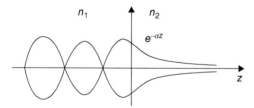

n_1 n_2 $e^{-\alpha z}$ z

Figure 2.6 Electric component of optical wave at the boundary of two media forming standing wave due to reflection, and wave decay $\sim e^{-\alpha z}$ due to refraction.

where the refractive index of the first transparent medium is larger than that of the second transparent medium, that is, $n_1 > n_2$, and when total reflection from the intersection occurs, that is, for incident angle exceeding the critical one, θ_c, defined by Eq. (2.26).

Even though the wave is propagating in the second medium, it transports no light energy in a direction normal to the boundary. All the light is totally internally reflected (TIR) at the boundary.

The guiding effect, which occurs in fiber optic structures (see Chapter 9), is based on TIR phenomenon:

> All energy transport occurs along the boundary of two media after TIR, without any penetration of light energy inside the intersection.

Moreover, we should notice that the totally internally reflected (TIR) wave undergoes a phase change, which depends on both the angle of incidence and the field polarization [15–17] (see also Chapters 9 and 10).

2.6 Propagation of Optical Waves in Material Media

As was shown above, each electrical field component (let say x-component) of the optical wave can be presented as a plane wave in any media in the following manner

$$\tilde{E}_x = Ae^{-\gamma z} + Be^{+\gamma z} \tag{2.31}$$

where A and B are constants, which can be obtained from the corresponding boundary conditions; the propagation parameter is complex and can be written as

$$\gamma = \alpha + j\beta \tag{2.32}$$

Here, α describes the attenuation of optical wave amplitude, that is, the wave energy losses, and β describes the phase velocity of the plane wave in the material media.

Now, we can present the magnetic field phasor component in the same manner as the electric field using (2.31)

$$\tilde{H}_y = \frac{1}{\eta}(Ae^{-\gamma z} - Be^{+\gamma z}) \tag{2.33}$$

where η is the intrinsic impedance of the medium, which is also complex. Solutions (2.31) and (2.33) can be concretized by using the corresponding boundary conditions. However, this is not the goal of our future analysis. We show the reader how the properties of the material medium change propagation conditions within it. For this purpose, we discuss further propagation parameters γ

(or α and β) and η associated with plane waves (2.31) and (2.33). After straight-forward computations of the corresponding equations, following [11–13, 17], we can find for $\mu = 1$ that

$$\alpha = \frac{\omega\sqrt{2\varepsilon}}{2}\left[\sqrt{1+\left(\frac{\sigma}{\omega\varepsilon}\right)^2}-1\right]^{1/2} \tag{2.34}$$

$$\beta = \frac{\omega\sqrt{2\varepsilon}}{2}\left[\sqrt{1+\left(\frac{\sigma}{\omega\varepsilon}\right)^2}+1\right]^{1/2} \tag{2.35}$$

The phase velocity is described by the propagation parameter β along the direction of propagation, which is defined by (2.35):

$$v_{\text{ph}} = \frac{\omega}{\beta} = \frac{\sqrt{2}}{\sqrt{\varepsilon}}\left[\sqrt{1+\left(\frac{\sigma}{\omega\varepsilon}\right)^2}+1\right]^{-1/2} \tag{2.36}$$

The dispersion properties follow from dependence on the frequency of the wave phase velocity $v_{\text{ph}} = v_{\text{ph}}(\omega)$. Thus, waves with different frequencies $\omega = 2\pi f$ travel with different phase velocities. In the same manner, the wavelength in the medium is dependent on the frequency of the optical wave:

$$\lambda = \frac{2\pi}{\beta} = \frac{\sqrt{2}}{f\sqrt{\varepsilon}}\left[\sqrt{1+\left(\frac{\sigma}{\omega\varepsilon}\right)^2}+1\right]^{-1/2} \tag{2.37}$$

In view of the attenuation of the wave with distance, the field variations with distances are not pure sinusoidal, as in free space. In other words, the wavelength is not exactly equal to the distance between two consecutive positive (or negative) extrema. It is equal to the distance between two alternative zero crossings.

We can now present formulas (2.34) and (2.35) using the general presentation of ε in the complex form, that is, $\varepsilon = \varepsilon' - j\varepsilon''$. If so,

$$\gamma^2 = (\alpha + j\beta)^2 = j\omega\mu(\sigma + j\omega\varepsilon'') - \omega^2\mu\varepsilon' \tag{2.38}$$

where now

$$\alpha = \frac{\omega\sqrt{2\varepsilon'}}{2}\left[\sqrt{1+\left(\frac{\sigma+\omega\varepsilon''}{\omega\varepsilon'}\right)^2}-1\right]^{1/2} \tag{2.39}$$

and

$$\beta = \frac{\omega\sqrt{2\varepsilon'}}{2}\left[\sqrt{1+\left(\frac{\sigma+\omega\varepsilon''}{\omega\varepsilon'}\right)^2}+1\right]^{1/2} \tag{2.40}$$

From the general formulas (2.39) and (2.40) some special cases follow for different kinds of material media.

2.6.1 Imperfect Dielectric Medium

This medium is characterized by $\sigma \neq 0$, but $\sigma/\omega\varepsilon \ll 1$. Using the following expansion

$$(1+x)^m = 1 + mx + \frac{m(m-1)}{2!}x^2 + \cdots$$

we can easily obtain from (2.39) and (2.40)

$$\alpha \approx \sqrt{\frac{1}{\varepsilon'} \frac{\omega\varepsilon''}{2}} \tag{2.41a}$$

$$\beta \approx \omega\sqrt{\varepsilon'}\left(1 + \frac{\varepsilon''^2}{8\varepsilon'^2}\right) \tag{2.41b}$$

Now, as was done from the beginning, we will introduce the complex refractive index $n = n' - jn''$ in the above expressions instead of permittivity, ε, where now $n' = \sqrt{\varepsilon'/\varepsilon_0}$ and $n'' = \sqrt{\varepsilon''/\varepsilon_0}$ [2, 3]. Then, we will get in the case of a low-loss dielectric (or "imperfect" dielectric)

$$\alpha \approx \frac{\omega n''}{c}, \quad \beta \approx \frac{\omega}{c}\sqrt{n'}\left(1 + \frac{n''^2}{8n'^2}\right), \quad \text{and} \quad n'' \approx n'\frac{\varepsilon''}{2\varepsilon'} \tag{2.42}$$

2.6.2 Good Conductor Medium

Good conductors are characterized by $\sigma/\omega\varepsilon \gg 1$, the opposite of imperfect dielectrics. In this case, $|\,\mathbf{j}_c\,| \sim \sigma\tilde{E}_x \gg |\,\mathbf{j}_d\,| \sim \omega\varepsilon\tilde{E}_x$; from (2.41a) and (2.41b) we obtain

$$\alpha \approx \frac{\omega n''}{c} \approx \sqrt{\frac{\sigma\omega}{2}} \quad \text{and} \quad n'' \approx \sqrt{\frac{\sigma}{2\omega\varepsilon}} \tag{2.43}$$

For example, if the absorption coefficient of glass at $\lambda = 10\,\mu m$ equals $\alpha = 1.8\,\text{cm}^{-1}$, then using (2.43), we finally get

$$n'' \approx \frac{\lambda}{2\pi}\alpha = \frac{10^{-5}\,\text{m} \cdot 1.8 \cdot 10^2\,\text{m}^{-1}}{2\pi} = 2.9 \cdot 10^{-4}$$

Problems

2.1 The angle of incidence of a plane EM wave at the boundary of two media is θ_0. In addition, $\varepsilon_{r1} = \varepsilon_{r2} = 3$ and $\mu_1 = \mu_2 = \mu_0$. The electric field of the incident wave is E_1 (V/m).

Find: (1) The angle of refraction θ_2. (2) The amplitude E_2 of the refracted wave.

2.2 An optical wave is normally incident upon the boundary of two materials with the relative parameters
$$\varepsilon_{r1} = 1, \quad \varepsilon_{r2} = 5, \quad \text{and} \quad \mu_{r1} = \mu_{r2} = 1$$
The amplitude of the electric field component of the incident wave field is $E_1 = 5$ (V/m).
Find: Coefficients of reflection and refraction and the amplitudes of the corresponding reflected and refracted wave.

2.3 A plane EM wave is incident on the boundary of two media with an angle of $\theta_0 = 60°$. The parameters of the media are $\varepsilon_{r1} = 1$, $\varepsilon_{r2} = 3$, and $\mu_{r1} = \mu_{r2} = 1$. The amplitude of electric field of the wave is $|E_0| = 3$ (V/m).
Find: (1) The coefficients of reflection and refraction for both types of wave polarization. (2) The corresponding amplitudes of the reflected and refracted waves. (3) Check your answer by making use of the relevant laws.

2.4 Rewrite formulas (2.19) and (2.20) in terms of the angle of refraction rather than the angle of incidence.

2.5 An optical wave with an amplitude of 1 (V/m) is normally incident upon the boundary between a dielectric having parameters $\varepsilon_1 = 4\varepsilon_0$ and $\mu_1 = \mu_0$ and air with parameters $\varepsilon_1 = \varepsilon_0$ and $\mu_1 = \mu_0$.
Find: The amplitude of the reflected and the transmitted fields and the incident, reflected, and transmitted power.

2.6 An optical wave with vertical polarization propagating through a dielectric medium is incident upon the boundary separating air and the dielectric at an angle of $\theta_0 = 60°$. The air and the dielectric have the following parameters: $\varepsilon_{r1} = 1, \mu_{r1} = 1$, and $\varepsilon_{r2} = 8, \mu_{r2} = 1$, respectively.
Find: Brewster's angle and the critical angle. Find these angles, if, conversely, the wave is propagating into the air from the dielectric material at the same angle of incidence $\theta_0 = 60°$.

2.7 Given are two media, air with the parameters $\varepsilon_1 = \varepsilon_0$, $\mu_1 = \mu_0$, $\eta_1 = \eta_0$ and glass with the parameters $\varepsilon_2 = 5\varepsilon_0$, $\mu_2 = \mu_0$, $\eta_2 = 150\,\Omega$. For both media the conductivities are $\sigma_1 = \sigma_2 = 0$. The optical one-dimensional (1D) wave with horizontal polarization having the form $\mathbf{E}_{0x} = 10 \exp\{i(10^8 \pi \cdot t - \beta_0 \cdot z)\}\mathbf{u}_x$ (V/m) is normally incident upon the boundary of two media.
Find: The parameter of wave propagation in the second media, β, the coefficients of reflection and refraction, and the amplitude of the reflected and the transmitted field from air to glass.

2.8 A vertically polarized optical wave with amplitude $|E_0| = 50\,(V/m)$ and zero initial phase is normally incident upon the boundary of two media: air ($\varepsilon_1 = \varepsilon_0$, $\mu_1 = \mu_0$, $\eta_1 = \eta_0$) and water ($\varepsilon_2 = 81\varepsilon_0$, $\mu_2 = \mu_0$, $\eta_2 \approx 42\,\Omega$). *Find*: The electric and magnetic components of the reflected and the transmitted waves via the corresponding coefficients of reflection and refraction.

2.9 A horizontally polarized optical wave is normally incident upon the boundary between air ($\varepsilon_1 = \varepsilon_0$, $\mu_1 = \mu_0$, $\sigma_1 = 0$) and a dielectric ($\varepsilon_2 = 81\varepsilon_0$, $\mu_2 = \mu_0$, $\sigma_2 = 0$). The field is given by the following expression:
$E_0(x, z) = [4\mathbf{u}_x - 3\mathbf{u}_z]\exp\{-j(6x + 8z)\}\,(V/m)$
Find: The frequency and the wavelength of this wave in the dielectric; the coefficient of reflection and the corresponding expression of the reflected field; the coefficient of refraction and the corresponding expression of the transmitted field.

2.10 A vertically polarized optical wave is incident upon the boundary between air ($\varepsilon_1 = \varepsilon_0$, $\mu_1 = \mu_0$) and a dielectric ($\varepsilon_2 = 6\varepsilon_0$, $\mu_2 = \mu_0$) at an angle of θ_0. The wave is described by $E_0(x, z) = 5[\cos\theta_0\mathbf{u}_x - \sin\theta_0\mathbf{u}_y]$ (V/m). The wavelength in air $\lambda_1 = 3\,cm$.
Find: Brewster's angle and the critical angle; the field expression after reflection assuming an incident angle that equals Brewster's angle; the field expression for the transmitted wave assuming an incident angle that equals Brewster's angle.

2.11 Find expressions for the skin depth and intrinsic impedance vs. frequency, f, for copper with conductivity $\sigma = 5.8\cdot 10^7\,S/m$ and relative permeability $\mu_r = 1$ (a good conductor).

2.12 Given: an e/m plane wave with a magnetic component having field strength $B_0 = 10\,A/m$ and frequency $f = 600\,kHz$ propagating in the positive direction along z-axis in copper with $\mu_r = 1$ and $\sigma = 5.8\cdot 10^7\,S/m$.
Find: α, β, the critical depth, δ, and $B(z, t)$.

2.13 In sea water with $\sigma = 4\,S/m$, $\varepsilon = 81\varepsilon_0$, the frequency of an optical wave is 100 THz. Find α and the attenuation (in dB/m) considering that the transmission of the wave is proportional to $\tau = \exp(-\alpha z)$. What will change if the frequency is decreased to 10 THz?

2.14 The absorption coefficient of glass at $\lambda = 10\,\mu m$ is $\alpha = 1.8\,cm^{-1}$.
Find: The imaginary part of the refractive index.

References

1 Akhamov, S.A. and Khohlov, R.V. (1965). *Problems of Nonlinear Optics.* Moscow: Fizmatgiz (in Russian).

2 Lipson, S.G. and Lipson, H. (1969). *Optical Physics.* Cambridge: University Press.

3 Akhamov, S.A., Khohlov, R.V., and Sukhorukov, A.P. (1972). *Laser Handbook.* North Holland: Elsevier.

4 Marcuse, O. (1972). *Light Transmission Optics.* New York: Van Nostrand-Reinhold Publisher.

5 Kapany, N.S. and Burke, J.J. (1972). *Optical Waveguides,* Chapter 3. New York: Academic Press.

6 Fowles, G.R. (1975). *Introduction in Modern Optics.* New York: Holt, Rinehart, and Winston Publishers.

7 Jenkis, F.A. and White, H.E. (1953). *Fundamentals of Optics.* New York: McGraw-Hill.

8 Born, M. and Wolf, E. (1964). *Principles in Optics.* New York: Pergamon Press.

9 Fain, V.N. and Hanin, Y.N. (1965). *Quantum Radiophysics.* Moscow: Sov. Radio (in Russian).

10 Grant, I.S. and Phillips, W.R. (1975). *Electromagnetism.* New York: Wiley.

11 Plonus, M.A. (1978). *Applied Electromagnetics.* New York: McGraw-Hill.

12 Kong, J.A. (1986). *Electromagnetic Wave Theory.* New York: Wiley.

13 Elliott, R.S. (1993). *Electromagnetics: History, Theory, and Applications.* New York: IEEE Press.

14 Kopeika, N.S. (1998). *A System Engineering Approach to Imaging.* Washington: SPIE Optical Engineering Press.

15 Dakin, J. and Culshaw, B. (eds.) (1988). *Optical Fiber Sensors: Principles and Components,* vol. 1. Boston-London: Artech House.

16 Palais, J.C. (2006). Optical Communications. In: *Engineering Electromagnetics Applications* (ed. R. Bansal). New York: Taylor & Francis.

17 Blaunstein, N. and Christodoulou, C. (2007). *Radio Propagation and Adaptive Antennas for Wireless Communication Links: Terrestrial, Atmospheric and Ionospheric.* New Jersey: Wiley InterScience.

Part II

Fundamentals of Optical Communication

3

Types of Signals in Optical Communication Channels

3.1 Types of Optical Signals

In optical wired or wireless links the same kinds of signal are formed and transmitted as in similar radio wired and wireless communication channels. They are continuous and discrete (e.g. pulses). Therefore, the same mathematical tool can be used for description of such kinds of signals, radio and optical. Let us briefly present a mathematical description of both types of signals – continuous wave (CW) and pulses. In communications, wired and wireless, there are other definitions of these kinds of signals that researchers use regarding their presentation in the frequency domain. Thus, if we deal with continuous signal in the time domain, let us say, $x(t) = A(t) \cos \omega t$, which occupies a wide time range along the time axis, its Fourier transform $F[x(t)]$ converts this signal into the narrowband signal, that is, $F[x(t)] = y(f)$, which occupies a very narrow frequency band in the frequency domain. Conversely, if we deal initially with pulse signal in the time domain that occupies a very narrow time range along the time axis, its Fourier transform $F[x(t)]$ converts this signal into the wideband signal, that is, $F[x(t)] = X(f)$, which occupies a huge frequency band in the frequency domain. Therefore, in communication terminology continuous signals and pulses are often called the *narrowband* and *wideband*, respectively. In our description below we will follow both terminologies in places where usage of different definitions is more suitable.

3.1.1 Narrowband Optical Signals

A voice-modulated CW signal occupies a very narrow bandwidth surrounding the carrier frequency f_c of the signal (e.g. the carrier), which can be expressed as

$$x(t) = A(t) \cos \left[2\pi f_c t + \phi(t) \right] \tag{3.1}$$

Fiber Optic and Atmospheric Optical Communication, First Edition.
Nathan Blaunstein, Shlomo Engelberg, Evgenii Krouk, and Mikhail Sergeev.
© 2020 John Wiley & Sons, Inc. Published 2020 by John Wiley & Sons, Inc.

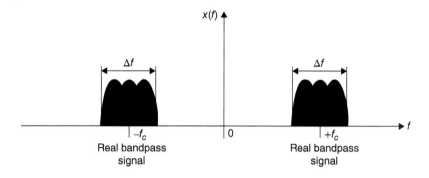

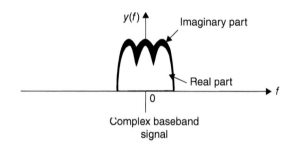

Figure 3.1 Comparison between baseband and bandpass signals.

where $A(t)$ is the signal envelope (i.e. slowly varied amplitude) and $\phi(t)$ is its signal phase. Since all information in the signal is contained within the phase and envelope-time variations, an alternative form of a bandpass signal $x(t)$ is introduced [1–8]:

$$y(t) = A(t)\exp\{j\phi(t)\} \tag{3.2}$$

which is also called the *complex baseband* representation of $x(t)$. By comparing (3.1) and (3.2), we can see that the relation between the *bandpass* and the *complex baseband* signals are related by

$$x(t) = \mathrm{Re}[y(t)\exp(j2\pi f_c t)] \tag{3.3}$$

Relations between these two representations of the narrowband signal in the frequency domain are shown schematically in Figure 3.1.

One can see that the complex baseband signal is a frequency shifted version of the bandpass signal with the same spectral shape, but centered on a zero frequency instead of the f_c [4–8]. Here, $X(f)$ and $Y(f)$ are the Fourier transform of $x(t)$ and $y(t)$, respectively, and can be presented in the following manner [1, 9–13]:

$$Y(f) = \int_{-\infty}^{\infty} y(t)e^{-j2\pi ft}dt = \text{Re}[Y(f)] + j\,\text{Im}[Y(f)] \tag{3.4}$$

and

$$X(f) = \int_{-\infty}^{\infty} x(t)e^{-j2\pi ft}dt = \text{Re}[X(f)] + j\,\text{Im}[X(f)] \tag{3.5}$$

Substituting for $x(t)$ in integral (3.5) from (3.3) gives

$$X(f) = \int_{-\infty}^{\infty} \text{Re}[y(t)e^{j2\pi f_c t}]e^{-j2\pi ft}dt \tag{3.6}$$

Taking into account that the real part of any arbitrary complex variable w can be presented as

$$\text{Re}[w] = \frac{1}{2}[w + w^*]$$

where w^* is the complex conjugate, we can rewrite (3.5) in the following form:

$$X(f) = \frac{1}{2}\int_{-\infty}^{\infty} [y(t)e^{j2\pi f_c t} + y^*(t)e^{-j2\pi f_c t}] \cdot e^{-j2\pi ft}dt \tag{3.7}$$

After comparing expressions (3.4) and (3.7), we get

$$X(f) = \frac{1}{2}[Y(f - f_c) + Y^*(-f - f_c)] \tag{3.8}$$

In other words, the spectrum of the real bandpass signal $x(t)$ can be represented by the real part of that for the complex baseband signal $y(t)$ with a shift of $\pm f_c$ along the frequency axis. It is clear that the baseband signal has its frequency content centered on the "zero" frequency value.

Now we notice that the mean power of the baseband signal $y(t)$ gives the same result as the mean square value of the real bandpass signal $x(t)$, that is,

$$\langle P_y(t)\rangle = \frac{\langle |y(t)|^2\rangle}{2} = \frac{\langle y(t)y^*(t)\rangle}{2} \equiv \langle P_x(t)\rangle \tag{3.9}$$

The complex envelope $y(t)$ of the received narrowband signal can be expressed according to (3.2) and (3.3), within the multipath wireless channel, as a sum of phases of N baseband individual multiray components arriving at the detector with their corresponding time delay, τ_i, $i = 0, 1, 2, \ldots, N - 1$ [4–8]:

$$y(t) = \sum_{i=0}^{N-1} u_i(t) = \sum_{i=0}^{N-1} A_i(t)\exp[j\phi_i(t, \tau_i)] \tag{3.10}$$

If we assume that during subscriber movements through the local area of service, the amplitude A_i time variations are small enough, whereas phases ϕ_i vary greatly due to changes in propagation distance between the source and the desired detector, then there are great random oscillations of the total signal $y(t)$ at the detector during its movement over a small distance. Since $y(t)$ is the phase sum in (3.10) of the individual multipath components, the instantaneous phases of the multipath components result in large fluctuations, that is, fast fading, in the CW signal. The average received power for such a signal over a local area of service can be presented according to Refs. [1–8] as

$$\langle P_{CW} \rangle \approx \sum_{i=0}^{N-1} \langle A_i^2 \rangle + 2 \sum_{i=0}^{N-1} \sum_{i,j \neq i} \langle A_i A_j \rangle \langle \cos[\phi_i - \phi_j] \rangle \tag{3.11}$$

3.1.2 Wideband Optical Signals

The typical *wideband* or *impulse* signal passing through the multipath communication channel shown schematically in Figure 3.2a follows from Refs. [2–8].

If we divide the time delay axis into equal segments, usually called bins, then there will be a number of received signals, in the form of vectors or delta functions. Each bin corresponds to a different path whose time of arrival is within

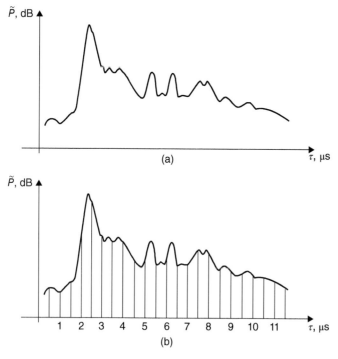

Figure 3.2 (a) A typical impulse signal passing through a multipath communication channel according to [1, 2, 9, 10]. (b) The use of bins as vectors for the impulse signal with spreading.

the bin duration, as depicted in Figure 3.2b. In this case, the time-varying discrete time impulse response can be expressed as

$$h(t, \tau) = \left\{ \sum_{i=0}^{N-1} A_i(t, \tau) \exp[-j2\pi f_c \tau_i(t)] \delta(\tau - \tau_i(t)) \right\} \exp[-j\phi(t, \tau)]$$

(3.12)

If the channel impulse response is assumed to be time invariant, or is at least stationary over a short time interval or over a small-scale displacement of the detector or source, then the impulse response (3.12) reduces to

$$h(t, \tau) = \sum_{i=0}^{N-1} A_i(\tau) \exp[-j\theta_i] \delta(\tau - \tau_i)$$

(3.13)

where $\theta_i = 2\pi f_c \tau_i + \phi(\tau)$. If so, the received power delay profile for a wideband or pulsed signal averaged over a small area can be presented simply as a sum of the powers of the individual multipath components, where each component has a random amplitude and phase at any time, that is,

$$\langle P_{\text{pulse}} \rangle = \left\langle \sum_{i=0}^{N-1} \{A_i(\tau) | \exp[-j\theta_i] | \}^2 \right\rangle \approx \sum_{i=0}^{N-1} \langle A_i^2 \rangle$$

(3.14)

The received power of the wideband or pulse signal does not fluctuate significantly when the subscriber moves within a local area, because in practice, the amplitudes of the individual multipath components do not change widely in a local area of service.

A comparison between small-scale presentations of the average power of the narrowband (CW) and wideband (pulse) signals, that is, (3.11) and (3.14), shows that

When $\langle A_i A_j \rangle = 0$ or/and $\langle \cos[\phi_i - \phi_j] \rangle = 0$, the average power for CW signal and that for pulse are equivalent.

This can occur when the path amplitudes are uncorrelated, that is, each multipath component is independent after multiple reflections, diffractions, and scattering from obstructions surrounding both the detector and the source. It can also occur when multipath phases are independently and uniformly distributed over the range of $[0, 2\pi]$. This property is correct for optical wavebands when the multipath components traverse differential paths having hundreds and thousands of wavelengths [4–8].

3.2 Mathematical Description of Narrowband Signals

Accounting for the identical description of radio and optical waves (e.g. rays) discussed in Chapter 2, we can state that all mathematical descriptions of radio waves are correct for the description of optical rays. We present below a canonical statistical description of optical signals, mostly mentioned in the literature of radio and optical communications [2–12, 14].

The Clarke 2-D model is a more suitable statistical model for satisfactory description of multipath ray phenomena in the optical communication channels, where the source and the detector heights are so low that the ray distribution in the vertical plane can be ignored [9].

The model assumes a fixed source and a moving receiver. The signal at the receiver is assumed to comprise of N horizontally traveling rays with each light ray with number i having equal average amplitude A_i and with statistically azimuthally independent angles of arrival (α_i) and signal phase (ϕ_i) distributions.

The assumption of equal average amplitude of each i-th wave is based on the absence of an line-of-sight (LOS) component with respect to scattered components arriving at the detector. Moreover, phase angles distribution is assumed to be uniform in the interval $[0, 2\pi]$, that is, the angle distribution function is equal, $P(\phi_i) = (2\pi)^{-1}$.

The receiver moves with a velocity v in the x-direction, so the Doppler shift in z-axis can now be rewritten as

$$f_D = \frac{v}{\lambda} \cos \alpha_i \tag{3.15}$$

The mean square value of the amplitude A_i of such uniformly distributed individual optical rays is constant:

$$E\{A_i^2\} \equiv \langle A_i^2 \rangle = \frac{E_0}{N} \tag{3.16}$$

because $N = $ const. and the real amplitude of local average field E_0 is also assumed to be a constant.

Let us consider the plane optical waves arriving at the moving detector, which usually has one E-field component (let say, the E_z) and two H-field components (H_x and H_y). Without any loss of generality of the problem, because for each field component the same technique is used, let us consider only the E-field component and present it at the receiving point as [9]

$$E_z = E_0 \sum_{i=1}^{N} A_i \cos(\omega_c t + \theta_i) \tag{3.17}$$

where $\omega_c = 2\pi f_c$, f_c is the carrier frequency; $\theta_i = \omega_i t + \phi_i$ is the random phase of the ith arriving component of total signal, and $\omega_i = 2\pi f_i$ represents the Doppler shift experienced by the ith individual wave. The amplitudes of all three components are normalized such that the ensemble average of the amplitude A_i is given by $\sum_{i=1}^{N} \langle A_i^2 \rangle = 1$ [9].

Since the Doppler shift is small with respect to the carrier frequency, all field components may be modeled as narrowband random processes and approximated as Gaussian random variables, if $N \to \infty$, with a uniform phase distribution in the interval $[0, 2\pi]$. If so, the E-field component can be expressed in the following form [9]:

$$E_z = C(t) \cos(\omega_c t) - S(t) \sin(\omega_c t) \tag{3.18}$$

where $C(t)$ and $S(t)$ are the in-phase and quadrature components that would be detected by a suitable receiver:

$$C(t) = \sum_{i=1}^{N} A_i \cos(\omega_i t + \theta_i)$$

$$S(t) = \sum_{i=1}^{N} A_i \sin(\omega_i t + \theta_i) \tag{3.19}$$

According to the assumptions above, both components $C(t)$ and $S(t)$ are independent Gaussian random processes that are completely characterized by their mean value and autocorrelation function [2–8, 11–14]. Moreover, they are also uncorrelated zero mean Gaussian random variables, that is,

$$\langle S \rangle = \langle C \rangle = \langle E_z \rangle = 0 \tag{3.20a}$$

with an equal variance σ^2 (the mean signal power) given by

$$\sigma^2 \equiv \langle |E_z|^2 \rangle = \langle S^2 \rangle = \langle C^2 \rangle = E_0^2/2 \tag{3.20b}$$

The envelope of the received E-field component can be presented as

$$|E(t)| = \sqrt{S^2(t) + C^2(t)} = r(t) \tag{3.21}$$

Since components $C(t)$ and $S(t)$ are independent Gaussian random variables that satisfy (3.20a)–(3.20b), the random received signal envelope r has a Rayleigh distribution [2–8, 11–14].

Hence, using the very simple two-dimensional model presented by Clarke [9], one can prove that it is possible to use in this case the well-known Rayleigh and Rician distributions to describe the fast fading usually observed in mobile optical and radio communication links [1–14].

In narrowband optical communication links most of the energy of the received signal is concentrated near the carrier frequency, f_c, and the power spectra is defined as the optical spectra of such a narrowband signal. At the same time, the real narrowband signal can be converted into the complex baseband signal presentation that has its frequency content centered on zero frequency. Below, we will describe narrowband fast fading through the power baseband spectra distribution in the space and time domains. For this purpose, we introduce the autocorrelation function $K(t, t + \tau) = \langle E(t) \cdot E(t + \tau) \rangle$ of signal fading in the time domain [2–5, 7, 8, 11–14]. It can be done by the use of, for example, the E-field components presentation according to [4, 7, 10, 12] in terms of a time delay τ, as

$$\langle E(t) \cdot E(t+\tau) \rangle = \langle C(t) \cdot C(t+\tau) \rangle \cos \omega_c \tau - \langle S(t) \cdot S(t+\tau) \rangle \sin \omega_c \tau$$
$$= c(\tau) \cos \omega_c \tau - s(\tau) \sin \omega_c \tau \qquad (3.22)$$

Here, operator $\langle \cdot \rangle$ describes the procedure of averaging, so we can rewrite the mean square value of amplitude A_i, introduced earlier by (3.19), as $E[A_i^2] \equiv \langle A_i^2 \rangle = E_0/N$. According to definition (3.20b), the correlation properties of signal fading are fully described by functions $c(\tau)$ and $s(\tau)$ [4, 7, 10, 12]:

$$c(\tau) = \frac{E_0}{2} \langle \cos \omega_c \tau \rangle, \quad s(\tau) = \frac{E_0}{2} \langle \sin \omega_c \tau \rangle \qquad (3.23)$$

For the particular case of Clarke's 2D model, where PDF(β) is the delta function, we get from (3.23)

$$c_0(\tau) = \frac{E_0}{2} I_0(2\pi f_m \tau) \qquad (3.24)$$

As is well known [8], the power spectrum of the resulting received signal can be obtained as the Fourier transform of the temporal autocorrelation function (3.24). Taking the Fourier transform of the component for the autocorrelation function (3.24), we finally, according to Clarke's 2D model, have the corresponding complex baseband power spectrum in the frequency domain:

$$W_0(f) = \frac{E_0}{4\pi f_m} \frac{1}{\sqrt{1 - \left(\frac{f}{f_m}\right)^2}}, \quad |f| \leq f_m$$

$$W_0(f) = 0, \quad \text{elsewhere} \qquad (3.25)$$

As follows from (3.25), the 2D spectrum is strictly band-limited within the range $|f| \leq f_m$ but the power spectral density becomes infinite at $f = \pm f_m$.

Now, let us compare the spectra of the different field components, because only the E_z component has been presented so far. For Clarke's statistical 2D model [9], the baseband signal power spectrum of the E_z component is given by (3.25). A similar analysis, as above, leads to expressions for the baseband spectra $W_0(f)$ distribution in the frequency domain for the magnetic field components H_x and H_y:

$$W_0(f) = \frac{E_0}{4\pi f_m} \sqrt{1 - \left(\frac{f}{f_m}\right)^2} \quad \text{for } H_x$$

$$W_0(f) = \frac{E_0}{4\pi f_m} \frac{\left(\frac{f}{f_m}\right)^2}{\sqrt{1 - \left(\frac{f}{f_m}\right)^2}} \quad \text{for } H_y \qquad (3.26)$$

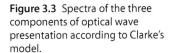

Figure 3.3 Spectra of the three components of optical wave presentation according to Clarke's model.

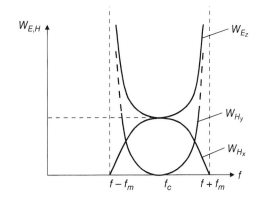

The form of these signal power spectra vs. the radiated frequency is presented in Figure 3.3 with the spectrum of the E_z component included for comparison.

As is shown in the illustrations, the E_z component, like the H_y component, has a minimum at the radiated carrier frequency and both are strictly limited to infinity at the range $\pm f_m$ around the carrier frequency while the H_x component has its maximum at the carrier frequency and is strictly limited to zero at the range $\pm f_m$.

We conclude therefore that the baseband signal spectrum of the resulting received signal is strictly band-limited to the range $\pm f_m$ around the carrier frequency, and within those limits the power spectral density depends on the probability density functions (PDFs) associated with the spatial angles of arrival α and β. The limits of the Doppler spectrum can be quite high. Frequency shifts of such a magnitude can cause interference with the message information.

3.3 Mathematical Description of Wideband Signals

The wideband or impulse response of the communication channel is schematically depicted in Figure 3.2a. If one now apportions the time delay axis into equal segments, usually called delay binds [2–5, 12], as shown in Figure 3.2b, then there will be a number of received signals in each bin corresponding to the different paths whose times of arrival are within the bin duration. If so, these signals can be represented vectorially using delta function presentation [1, 2, 9, 10] centered on each bin, as shown in Figure 3.2. This time-varying discrete time impulse response can be expressed by (3.12) in the general case, or by (3.13) in some specific case.

As is well known, the main factors that create the random signal modulation from the stationary or moving source are the multiray effects [1, 2, 7–12, 14]. All the time there is a random number of rays that arrive at the reception point and this number changes over time during the source's movements. The lifetime of each ray is determined by the time of the no-shadowing effect of the nontransparent screen scatterers with respect to both transmitter and receiver; this time is also a random measure. Different conditions of reflection from nontransparent reflector screens and variations of reflected rays' paths lead to random variations of signal amplitude and phase. At the same time, the movement of the radiation source in any direction leads to the Doppler shift of ray frequencies, an effect which depends on differences between directions to the reflectors and the direction of the source's movement. Taking into account all the abovementioned, let us now introduce the semi-width of spectrum $W(\omega)$ of wideband signal as

$$\Omega < \omega_0 \tag{3.27}$$

where ω_0 is the central (carrier) frequency of the optical signal spectrum.

According to condition (3.27) we can neglect the differences in Doppler shift for all spectral components and assume that the signal in the nth ray has the Doppler shift, using the current frequency ω_0 instead of ω_n, that is, $\omega_{dn} = \omega_0 \frac{v}{c} \cos \phi_n$. Moreover, in conditions of low-range light rays propagation, the propagation path is from tens to hundred meters, which is much higher compared to the wavelength of optical waves (see Chapter 1). In this case, together with in equation (3.27) we can assume the following conditions:

$$\omega_{dn} t_n \ll \pi, \quad \Omega t_n \ll \pi \tag{3.28}$$

and neglect the corresponding phase alterations. The radiated wideband signal that satisfies conditions (3.28) can be considered as a stationary random process with its energetic spectrum limited to the width $\Omega \pm \Delta \omega$ [11, 14]. The complex amplitude $A_s(t)$ of such a spectrum can be described in the time domain by the correlation function presented in the following form:

$$K_s(\tau) = \frac{\sigma^2}{\exp\left(-\frac{|\tau|}{\tau_k}\right)} \tag{3.29}$$

where τ_k is the correlation time. The energetic spectrum of the process $A_s(t)$, which corresponds to (3.29), can be easily found as

$$W_s(\omega) = \frac{2\sigma^2/\tau_n}{1 + (\tau_k \omega)^2} \tag{3.30}$$

where $\sigma^2 = \langle (\tau - \langle\tau\rangle)^2 \rangle$, $\tau_n = t_{2n} - t_{1n}$ is the lifetime of the nth ray, and $\langle \tau_n \rangle$ is the mean value. Using all the definitions mentioned above, we completely determine the signal $X(t)$ as a random process within the multipath wideband communication channel.

Let us now consider the signal characteristics of the nth ray. According to the discussions presented above, the receiving signal can be presented as a product of the radiated signal $s(t)$ and the transfer function $x(t, \omega_0)$ for the current oscillation; we can therefore consider the passage of the monochromatic signal with frequency ω_0 along the nth ray. The lifetime of such a ray is limited and therefore a signal passing along this ray will be received at the observed point as a pulse with duration τ_n, with complex amplitude a_n, and with the frequency shift ω_{dn} with respect to the current frequency ω_0.

As for the Doppler frequency shift, ω_{dn}, if we assume uniform distribution of reflectors in the arrival angle interval $[0, 2\pi]$, its probability distribution is proportional to $\sin^{-1}(\phi_n)$. But the random positions of scatterers in the angle-of-arrival domain lead to a very specific asymmetric form of the energetic spectrum of multi-scattered signal, which differs from the classical frequency distribution presented above (see Figure 3.3), with the infinite probability density at the boundaries of the band with width $2\omega_{dn\max} = 2\omega_0\frac{v}{c}$. We must also note that for the uniform and regular distribution of arrival angles, ϕ_n, within the range of $[0, 2\pi]$, the PDF of Doppler shift, ω_{dn}, when the source or the detector is not "shadowed" by the surrounding obstructions, with an accuracy of constant value, is described by the term proportional to that obtained by Clarke [9]:

$$\sim \left[\left(\omega_0 \frac{v}{c} \right)^2 - \omega^2 \right]^{-1/2} \tag{3.31}$$

As was analyzed in [12], the movements and the shadowing effect obstructions sharply decrease the area of the sources within which the scatterers are formed. Moreover, due to the screening effect of nontransparent screens, there is a gap in signal power distribution along the LOS line between the light source and the detector, when this line is intersected by these screens. For any orientation in space of these screens, they have shadowing effects on the detector. These effects can be understood only after obtaining the angle-of-arrival and frequency shift distributions of the PDF of the received signal. In this irregular case, we can state that the PDF of the Doppler shift frequency, ω_{dn}, may differ from the symmetric distribution (3.25), obtaining an asymmetric form with the same maximum at the boundaries of the frequency band $[-\omega_{d\max}, +\omega_{d\max}]$.

References

1 Marcuse, O. (1972). *Light Transmission Optics*. New York: Van Nostrand – Reinhold Publisher.

2 Jakes, W.C. (1974). *Microwave Mobile Communications*. New Jersey: IEEE Press.

3 Lee, W.Y.C. (1985). *Mobile Communication Engineering*. New York: McGraw Hill Publications.

4 Saunders, S.R. (1999). *Antennas and Propagation for Wireless Communication Systems*. New York: Wiley.

5 Feuerstein, M.L. and Rappaport, T.S. (1992). *Wireless Personal Communication*. Boston-London: Artech House.

6 Steele, R. (1992). *Mobile Radio Communication*. New Jersey: IEEE Press.

7 Rappaport, T.S. (1996). *Wireless Communications*. New York: Prentice Hall PTR.

8 Proakis, J.G. (1995). *Digital Communications*. New York: McGraw Hill.

9 Clarke, R.H. (1968). A statistical theory of mobile-radio reception. *Bell Syst. Tech. J.* 47: 957–1000.

10 Aulin, T. (1979). A modified model for the fading signal at a mobile radio channel. *IEEE Trans. Veh. Technol.* 28 (3): 182–203.

11 Stark, H. and Woods, J.W. (1994). *Probability, Random Processes, and Estimation Theory for Engineers*. New Jersey: Prentice Hall.

12 Blaunstein, N. (2004). Wireless communication systems, Chapter 12. In: *Handbook of Engineering Electromagnetics* (ed. R. Bansal). New York: Marcel Dekker.

13 Krouk, E. and Semionov, S. (eds.) (2011). *Modulation and Coding Techniques in Wireless Communications*. Chichester, England: Wiley.

14 Leon-Garcia, A. (1994). *Probability and Random Processes for Electrical Engineering*. New York: Addison-Wesley Publishing Company.

4

An Introduction to the Principles of Coding and Decoding of Discrete Signals

When dealing with digital modulation principles, one is presented with digital information in the form of a discrete stream of bits that represents different digital signals – often signals that have been encoded in some way. In this chapter, we list and describe the most important principles of discrete signal coding and decoding, principles that were developed in [1–42]. In Chapter 5, we make use of the theory presented here to analyze methods for transmitting a discrete information stream through optical communication channels, and in Chapter 12, some of the probabilistic characteristics of error detection in transmission via communication channels will be discussed. This chapter briefly summarizes results and discussions in [1–42], including those obtained by one of the authors with his colleagues [7, 38, 40].

4.1 Basic Concepts of Coding and Decoding

4.1.1 General Communication Scheme

The history of coding theory starts from Shannon's fundamental works. Results obtained since then in the framework of this theory by researchers all over the world are widely used in modern standards of wireless and wired communication and data storage systems [1, 8, 12, 38, 40–42].

Let us consider the communication scheme given in Figure 4.1. The sender needs to transmit information to the recipient, but information transmitted over the communication channel can be corrupted by the channel (by the physical environment, the various devices, the information processing procedures, etc. – see Chapters 1, 7, and 11). The coder and decoder must take the information to be transmitted and transform it in a way that allows us to detect errors and to try to correct them.

Let us define a *code* as a subset of all sequences consisting of n symbols where the symbols are taken from a given discrete alphabet. If the sequences to be

Fiber Optic and Atmospheric Optical Communication, First Edition.
Nathan Blaunstein, Shlomo Engelberg, Evgenii Krouk, and Mikhail Sergeev.
© 2020 John Wiley & Sons, Inc. Published 2020 by John Wiley & Sons, Inc.

Figure 4.1 Scheme of discrete (coded) communication.

encoded have k symbols from the same alphabet, then $k < n$. Let us call n the code's *length*, and let us call the ratio k/n the code's *rate*, R.

During the transfer, processing, and storage of data, code symbols are affected by noise. Noise characteristics depend on channel properties and are usually described as random process (see Chapter 6). It is worth mentioning that the noise sources are not necessarily stationary. It is customary to include all noise sources in the model of the "communication channel" shown in Figure 4.1. This includes any and all changes to the coded symbols starting from the moment the coder finished its work (and, usually, the modulator commences working) up to the moment the decoder starts working (generally after symbols are output by the demodulator).

The task of the decoder is to determine if noise in the channel changed the coded symbols (to detect an error) and, if an error is seen to have occurred, to try to correct it. Error detection usually involves checking whether the received word is part of the code, and error correction generally involves deciding which word was transmitted (using predetermined criteria). If the decoder chooses the wrong word, then there has been a *decoding error*. This event has a certain probability, and this probability is one of the most important characteristics of a communication system.

Shannon's theorems are fundamental results in coding theory (and, more generally, in information theory) [15]. These theorems were formulated for communication channels and show that channels of the type considered by Shannon can be characterized by the value C, called the *throughput capacity* or simply the *capacity* of the channel. This value depends on the properties of the channel [4, 10, 11].

The noisy channel coding theorem states that if the transmission rate does not exceed the capacity C of the channel, then there exist coding schemes that allow the information to be transmitted with a vanishingly small probability of error. That is, reliable communication is possible. On the other hand, if the transmission rate exceeds the capacity of the channel, the probability of error is bounded from below by a positive value: It is impossible to establish reliable communication at that rate.

Unfortunately, Shannon's theorem for a noisy channel is an existence theorem: It says that there are codes that allow reliable communication but does not describe a practical way of searching for such codes. There are classes of codes (e.g. low-density parity check codes – see Chapter 5) that approach the fundamental limits set by Shannon's theorem, but finding such codes remains a non-trivial problem.

In order to determine how well a communication system will perform, it is necessary to

(1) be able to describe the impact of the communication channel on the transmitted data, that is, to have a channel model – a mathematical description of the probability of different types of errors during transmission;
(2) have the coding and decoding algorithms;
(3) define the parameters of the desired code (generally influenced by the error probability).

In order to deal with (1), a model of the communication channel must be created. On the one hand, the model should be sufficiently simple to allow one to estimate relevant probabilities. On the other hand, it must be sufficiently accurate that decisions made on the basis of the model will be appropriate for the actual channel. When it is impossible to develop an analytical model of the channel, we sometimes turn to simulation models, which, in the final analysis, can be a rather complicated procedure and require a large sample size for small target error probability values. Below, we consider commonly used models of communication channels.

4.1.2 The Binary Symmetric Channel (BSC)

The binary symmetric channel (BSC) is one of the best known channel models in coding theory. A simple figure representing this model is given in Figure 4.2. The input and output alphabets of the channel are identical and consist of the two symbols 0 and 1. The probability of error does not depend on the value of the transmitted symbol and is denoted by p. From the figure, we understand that the probability of a 0 becoming a 1 or a 1 becoming a 0 is p.

The capacity of the BSC can be calculated by the use of the formula: $C_{BSC} = 1 - \eta(p)$, where $\eta(p) = -p\log_2 p - (1-p)\log_2(1-p)$ is known as the binary entropy function [2].

In the BSC, the probability of an error occurring is not dependent on whether an error or errors occurred previously. Such channels are said to be *memoryless*. For this reason, the probability of v errors over the course of n transmissions is

$$P(n, v) = C_n^v p^v (1 - p)^{n-v} \tag{4.1}$$

Figure 4.2 The binary symmetric channel.

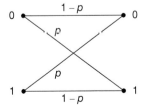

Equation (4.1) will be taken into account when the error probability of decoding procedures is analyzed. For many years, the BSC model was the basis for the construction of noise-tolerant codes. However, its simplicity does not allow it to adequately describe many real communication channels. The input and output alphabets of this model are discrete, whereas a more realistic model is a semi-continuous communication channel where the input alphabet is digital but the output alphabet is a continuous set of signals.

For many years, the basic semicontinuous channel model has been the Gaussian channel or the channel with additive white Gaussian noise (AWGN). Below, we consider a channel with AWGN and bipolar input based on the results obtained in [42].

4.1.3 Channel Model with AWGN

Let us suppose that the input alphabet is bipolar and consists of the symbols $x \in \{\pm 1\}$. We define the output symbol as $y = x + \eta$, where $\eta \in N(0, \sigma^2)$ is a normally distributed random value with zero mean and variance σ^2. The distribution of the signal at the output of the channel for inputs of $+1$ and -1 is shown in Figure 4.3. As with the BSC, the AWGN channel is a memoryless channel. The capacity of this channel can be shown to be

$$C_{\text{AWGN}} = \frac{1}{2} \sum_{x=\pm 1} \int_{-\infty}^{+\infty} p(y \mid x) \log \left(\frac{p(y \mid x)}{p(y)} \right) dy \qquad (4.2)$$

where the *conditional* probability in integral (4.2) is given by

$$p(y \mid x = \pm 1) = \frac{1}{\sqrt{2\pi}\sigma} \exp(-(y \mp 1)^2/(2\sigma^2)) \qquad (4.3)$$

Assuming that the input to the channel is $+1$ and -1 with equal probability, the relation between the *probability* $p(y)$ in integral (4.2) and the *conditional* probability given by (4.3) is given by:

$$p(y) = \frac{1}{2}(p(y \mid x = +1) + p(y \mid x = -1)) \qquad (4.4)$$

Just as the BSC is characterized by the probability of error p, the AWGN-model is characterized by the signal-to-noise ratio (SNR) E_b/N_0, where E_b is the energy of one transmitted bit and $N_0 = 2\sigma^2$ is the power density of the Gaussian process.

In the case of a bipolar input alphabet, $E_b = 1$. However, if we are transmitting R symbols per second, then E_b is the energy of one information bit and must be calculated as $E_b = 1/R$. After taking this into account, we find that $E_b/N_0 = 1/(2R\sigma^2)$. It is customary to report this value in decibels (dB).

If we estimate the input to the channel by considering $\hat{x} = \text{sign}(y)$ (that is, if we use a *hard-decision decoder*), then the AWGN channel with the hard-decision

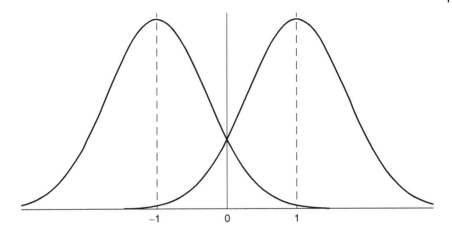

Figure 4.3 A semicontinuous Gaussian channel with bipolar input.

decoder can be modeled as a BSC channel with error probability

$$p = \int_{1/\sigma}^{\infty} \frac{1}{\sqrt{2\pi}} e^{-\beta^2/2} d\beta \qquad (4.5)$$

One of the principal differences between the AWGN channel and the BSC is the possibility of making use of both the sign and the magnitude of the received signal. The larger the magnitude of the received signal, the more likely the decoder decodes correctly (see Figure 4.3). This reliability can be quantified using the log likelihood ratio (LLR):

$$\text{LLR}(x \mid y) = \log \frac{\Pr(x = +1 \mid y)}{\Pr(x = -1 \mid y)} \qquad (4.6)$$

Decoders that make decisions about each symbol separately are said to be *hard-decision decoders*; decoders that make use of the information about sign and magnitude (as a rule, in the form of the LLR) of the received signals and consider several symbols together are called *soft-decision decoders*. Using a soft-decision decoder in the AWGN channel rather than a hard-decision decoder usually causes the decoding to improve to the same extent that increasing the SNR by 2–3 dB would.

4.2 Basic Aspects of Coding and Decoding

4.2.1 Criteria of Coding

The decoder must, somehow, compare the received word with the codewords. It is necessary to find criteria to be used in this comparison. If there is a sufficiently detailed channel model, then the natural way to associate a codeword with the received word is to associate the codeword that is most likely to have been transmitted given the received word. Let us formulate this criterion more formally.

Let the transmitted codeword be $x = (x_1, \ldots, x_n)$ and the received word be $\mathbf{y} = (y_1, \ldots, y_n)$. A channel can be described by the conditional probability $p(\mathbf{y}|\mathbf{x})$, and for channels without memory (such as the BSC or the AWGN channel), $p(\mathbf{y} \mid \mathbf{x}) = \prod_{i=1}^{n} p(y_i \mid x_i)$. This conditional probability function is called the *likelihood* function (LF). Given this function and the probability density function of the input, $p(\mathbf{x})$, we choose $\hat{x}$ to be the sequence that maximizes the probability of $\mathbf{y}$ having been received given that a sequence $\mathbf{x}$ was transmitted. This criterion is known as the *maximum a posteriori probability criterion* and can be expressed by

$$\hat{\mathbf{x}} = \arg\max_{\mathbf{x}} p(\mathbf{x} \mid \mathbf{y}) = \arg\max_{\mathbf{x}} \frac{p(\mathbf{y} \mid \mathbf{x})p(\mathbf{x})}{p(\mathbf{y})} \tag{4.7}$$

When all the codewords are equally likely (which is often true in practice), the *maximum a posteriori probability* can be replaced by the *maximum likelihood criterion*:

$$\hat{\mathbf{x}} = \arg\max_{\mathbf{x}} p(\mathbf{y} \mid \mathbf{x}) \tag{4.8}$$

However, the criteria introduced above requires the computation of the conditional probabilities (which are used to describe the channel), and, therefore, does not allow us to determine the requirements that must be placed on the code. That is, this criteria does not lead to a simple way of formulating the code-design problem. For this reason, decoding by using the *minimum distance* between the received word and one of the codewords is often used:

$$\hat{\mathbf{x}} = \arg\min_{\mathbf{x}} d(\mathbf{x}, \mathbf{y}) \tag{4.9}$$

Here $d(\mathbf{x}, \mathbf{y})$ is a distance function that makes it possible to determine "how close the vectors $\mathbf{x}$ and $\mathbf{y}$ are." The function $d(\mathbf{x}, \mathbf{y})$ should be taken to be a metric – a function $d(\mathbf{x}, \mathbf{y})$ having the following properties:

1. Positivity (nonnegativity): $d(\mathbf{x}, \mathbf{y}) \geq 0$, where $d(\mathbf{x}, \mathbf{y}) = 0 \Leftrightarrow \mathbf{x} = \mathbf{y}$
2. Symmetry: $d(\mathbf{x}, \mathbf{y}) = d(\mathbf{y}, \mathbf{x})$
3. Triangle Inequality: $d(\mathbf{x}, \mathbf{y}) \leq d(\mathbf{x}, \mathbf{z}) + d(\mathbf{z}, \mathbf{y})$.

Frequently, the Hamming distance, $d_H(\mathbf{x}, \mathbf{y})$, is used. The value of $d_H(\mathbf{x}, \mathbf{y})$ is equal to the number of positions in which $\mathbf{x}$ and $\mathbf{y}$ differ. In the discrete channel, the Hamming distance between the transmitted word and the received word is the number of mistakes that occurred during the transmission of the word $\mathbf{x}$. In [38, 42], it is shown that decoding by using the maximum of the log-likelihood function (LLF) is similar to decoding based on minimizing the Hamming distance between a valid codeword and the received word.

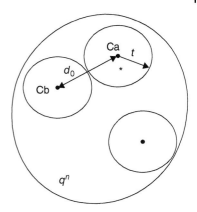

Figure 4.4 Geometric description of a t-error correcting code. Because the received n-tuple is located in the ball of radius t around the codeword Ca, it will be decoded as Ca.

Decoding continuous time transmissions by using the maximum of the LLF in an AWGN channel coincides with decoding using the minimum distance in the Euclidian metric. In what follows, we only consider the Hamming distance.

In Figure 4.4, a geometric description of a t-error correcting code is presented.

In Figure 4.4, the codewords are depicted by points in A^n. Adding a metric to this space allows one to compute the distance between any two points. Given such a metric, a *distance function*, we define d_0 as the minimal distance between any pair of codewords. We consider a sphere of radius $t = \lfloor (d_0 - 1)/2 \rfloor$ about each codeword. This is the largest radius for which no two spheres will have any common points. Decoding based on minimal distance is, thus, searching for the codeword nearest the received n-tuple of symbols (denoted in Figure 4.4 by an asterisk, *).

The decoding procedure will produce erroneous results if the received word was "closer" to an incorrect codeword than it was to the transmitted codeword. As long as fewer than t errors are introduced, the closest codeword will always be the transmitted codeword – and our decoding technique will work properly.

Once $t + 1$ symbols in the transmitted codeword are changed, it is possible that the codeword will not be properly decoded. (The corrupted codeword may "move" from the sphere around the correct codeword to the sphere around a different codeword.)

The minimal distance between two valid codewords is one of the parameters that determines how "good" a code is. The larger d_0 is, the more errors can be corrected by the code. In most practical coding and decoding schemes, one is generally told that the communication channel will not cause more than t errors to be introduced to a given transmission; when more than t errors are introduced, one may not be able to correct the errors [6–12].

The code distance provides a convenient way to describe the errors that the code can correct. If the number of errors over the length of the codeword is less than half the code distance, the errors can be corrected. Typically, the code can correct many more errors than half the code distance. This, however, usually requires an exponential number of operations.

We can say that the decoding criterion above, which makes use of the minimal distance between codewords, d_0, leads to *decoding within a sphere of radius t*, in which the codeword located in a sphere of radius t around the received word is "declared" to be the transmitted codeword. If the received word is not located in any of the spheres of radius t, the algorithm can either return "unable to decode" or return an arbitrary codeword [12].

This criterion is the basis for many important practical coding and decoding schemes such as cyclic codes [12].

4.2.2 Code Parameters for Error Correction

Any system of information transfer is characterized by the following parameters [38]:

1. P_{error} – error probability
2. V – rate of data transmission (bits per sec, *bps*)
3. ϑ – complexity of arrangement of coder/decoder.

When designing a communication system the designer aims to move data quickly with a small probability of error and using the lowest complexity coding/decoding scheme that is practical. The designer's goals can be represented by the following illustration in which ↑ denotes the increase of and ↓ denotes the decrease of the corresponding parameters. That is,

$$P_{error} \downarrow, \quad V \uparrow, \quad \vartheta \downarrow$$

In other words, the designer is required to minimize the error probability, increase the rate at which information is transmitted, and reduce the complexity of the coding/decoding scheme. As described above, the ability of a code of length n and rate R to correct errors (that is, to achieve a particular error probability) is often determined by the minimal distance d_0.

The characteristics of the code are related to those of the communication channel introduced above. The complexity ϑ generally depends on the code length n, and the rate of data transfer V depends on the code rate R, which is related to the number of words in the code, N, because $N = q^{nR}$. Also, the probability of a decoding error, P_{error}, generally decreases with increasing d_0.

It is, therefore, reasonable to design codes that allow us to achieve

$$d_0 \uparrow, \quad N \uparrow, \quad n \downarrow$$

The requirements above contradict one another. The code length n leads to an upper bound of q^n on the size of N, the number of possible codewords. If we fix n, growth of N inevitably leads to a decrease in d_0, and conversely, decreasing N leads to increase in d_0. Thus, the task of designing a good code can often be formulated as optimizing one of the parameters, d_0, N, n, while fixing the remaining parameters. For example, one might look to maximize $N(n, d_0)$ for fixed values of n and d_0.

In order to "measure" how good a code is, one often considers how close N comes to its maximum possible value for fixed values of n and d_0.

Upper bound on N: The Hamming bound [12] states that for any q-ary code of length n, with $d_0 = 2t + 1$ the following bound on N holds:

$$N \le \overline{N} = \frac{q^n}{\sum_{i=0}^{t} C_n^i (q-1)^i} \tag{4.10}$$

Codes for which (4.10) is an equality (for which $N = \overline{N}$) are said to be *perfect* codes. All perfect codes are known – (other than certain trivial perfect codes) they are either Hamming codes or Golay codes. There are no other perfect codes [12]. The Hamming bound guarantees that there are no codes for which $N > \overline{N}$.

A lower bound on N: The Gilbert–Varshamov bound for the important class of codes, called *linear* codes (which will be considered below), states that for a given q, n, and d_0, it is possible to find an N for which the following inequality holds:

$$N \ge \underline{N} = \frac{q^n}{\sum_{i=0}^{d_0-2} C_{n-1}^i (q-1)^i} \tag{4.11}$$

That is, there exists a q-ary linear code of length n with minimal distance d_0 and N codewords.

In other words, the Gilbert–Varshamov bound guarantees the existence of codes with a number of codewords that is very near the lower bound. Such codes are generally called "good," but finding such codes in the general case is still an open problem.

4.2.3   Linear Codes

In the previous sections, we considered the code as a mapping between a set of q^k information words of length k and codewords of length n. It is clear that other than for small values of k and n, it is impossible to realize the coder/decoder by making use of a list of all the codewords. (Working this way would require that every received word be compared with the list.) It is necessary to develop a simpler decoding procedure. It is possible to design such procedures by adding additional structure to the spaces we deal with. In this case, it is convenient to work with *linear vector spaces*.

A set V is called a *linear vector space* over the field K if

1. V has a commutative summation operation;
2. for any $\mathbf{v} \in V$, $\lambda \in K$ the product of the two elements is defined as $\lambda \mathbf{v} \in V$. (The product of a vector $\mathbf{v}$ and a scalar λ exists and is in V.) The following properties must also hold:
   a. $\lambda(\mathbf{v}_1 + \mathbf{v}_2) = \lambda \mathbf{v}_1 + \lambda \mathbf{v}_2$
   b. $(\lambda_1 + \lambda_2)\mathbf{v} = \lambda_1 \mathbf{v} + \lambda_2 \mathbf{v}$
   c. $(\lambda_1 \lambda_2)\mathbf{v} = \lambda_1(\lambda_2 \mathbf{v})$
   d. $1_K \mathbf{v} = \mathbf{v}$.
   Here 1_K is the neutral element of the product operation of the field K.

In what follows, we will generally consider vector spaces over the field $GF(2)$. In each linear space of dimension k there is a basis, that is, a set of k linearly independent vectors. The linear space consists of all 2^k linear combinations of basis vectors.

We will say that a code is a *linear binary (n, k)-code* if its elements form a k-dimensional subspace of the n-dimensional space of binary sequences.

Because (n, k)-codes are a k-dimensional subspace, they can be described using a basis of k linearly independent vectors. Let us "store" this basis in a matrix $\mathbf{G}$ of dimension $(k \times n)$. We will refer to this matrix as the *generator matrix* of the code G.

With $\mathbf{G}$ defined in this way, coding can be performed as follows. Let us suppose that $\mathbf{m}$ is an information vector of length k. Then the codeword $\mathbf{a}$ can be taken to be the product of the information vector and the generator matrix $\mathbf{G}$:

$$\mathbf{m} \overset{\varphi}{\to} \mathbf{a} = \mathbf{m}\mathbf{G}$$

Many of the important parameters of the code are immediately apparent when one looks at $\mathbf{G}$. It is easy to determine the code's length n, the number of codewords $N = 2^k$, and the rate $R = k/n$. Only one parameter is not easy to determine – the minimal distance, d_0.

Let us now represent $\mathbf{G}$ as follows: $\mathbf{G} = [\mathbf{C}_1 \,|\, \mathbf{C}_2]$, where $\mathbf{C}_1$ is a square $(k \times k)$-matrix and $\mathbf{C}_2$ is of dimension $(k \times n - k)$.

Then, if $\mathbf{C}_1$ is invertible, then, by multiplying $\mathbf{G}$ by $\mathbf{C}_1^{-1}$, we will obtain a different basis, $\mathbf{G}'$, of this linear space:

$$\mathbf{G}' = [\mathbf{I}_k \,|\, \mathbf{C}] \tag{4.12}$$

where $\mathbf{I}_k$ is the unit $(k \times k)$-matrix and $\mathbf{C} = \mathbf{C}_1^{-1}\mathbf{C}_2$. By using $\mathbf{G}'$ to encode data, we get

$$\mathbf{m} \overset{\varphi}{\longrightarrow} \mathbf{a} = \mathbf{m}\mathbf{G}' = (\ \underbrace{\mathbf{m}}_{k} \ |\ \underbrace{\mathbf{c}}_{n-k}\ )$$

Now, the first k positions of the codeword always consist of the original information vector. Codes of this sort – such as the code associated with the matrix that appears in (4.12) – are called *systematic codes*.

Let us consider the matrix

$$H = [-C^T \mid I_r] \tag{4.13}$$

of dimensions $(r \times n)$, where $r = n - k$ is the number of additional symbols and C, of dimension $(k \times r)$, is the matrix from (4.12). We note that for the matrices over $GF(2)$, $-C = C$. It can be shown that

$$GH^T = [I_k \mid C] \begin{bmatrix} -C \\ \hline I_r \end{bmatrix} = -C + C = 0$$

That is, the rows of the matrix in (4.13) are a basis of the space *orthogonal* to the code generated by G.

If the first k columns of the generator matrix G are linearly dependent (i.e. the matrix C_1 is singular), as the rows of G are known to be a basis, the rank of G must equal k, and, therefore, the generator matrix contains k linearly independent columns. Exchanging these columns with those in the first k positions in G, we are now able to proceed as we did above.

We notice that for any codeword a the following equation is valid:

$$aH^T = (mG)H^T = m(GH^T) = 0$$

On the other hand, for any vector e that is not a valid codeword, the product eH^T cannot be equal to the zero vector, because e does not lie in the span of the rows of G and cannot be orthogonal to all of the rows of H.

As a result, we find that a second way to define a linear code is to say that it is the set of words, $\{a\}$, for which $aH^T = 0$. That is, the matrix H also uniquely determines the linear code. This matrix is called the code's check matrix (as with the help of the matrix H it is easy to check whether an arbitrary vector is a codeword).

There is a simple procedure to detect errors. Let us suppose that the word b was received. By calculating the *syndrome*

$$S = bH^T \tag{4.14}$$

we can determine whether b is a valid codeword. If the syndrome equals zero, b is a valid codeword. Otherwise, an error occurred during transmission, and b is not a valid codeword.

With the help of the generator and check matrices, it is easy to produce codewords and detect errors in codewords from a linear code. As mentioned above, the error correction capabilities of a code are usually characterized by the minimal distance d_0 of the code. The minimal distance of a linear code is determined by the properties of the generator and check matrices.

Let $W(\mathbf{a})$ be the Hamming weight of an arbitrary vector $\mathbf{a}$. That is, let $W(\mathbf{a})$ be the number of nonzero elements in the codeword $\mathbf{a}$. We call the *minimal weight* W_0 of the code G the minimal Hamming weight of the nonzero codewords, $W_0 = \min_{\mathbf{a}_i \in G} W(\mathbf{a}_i)$.

In order for a linear (n, k)-code with check matrix $\mathbf{H}$ to have minimal distance d_0, it is necessary and sufficient that

1. any $d_0 - 1$ columns of $\mathbf{H}$ are linearly independent;
2. the matrix $\mathbf{H}$ contains a set of d_0 linearly dependent columns.

In order to calculate the minimal distance, it is sufficient to find the minimal set of linearly dependent columns of the check matrix $\mathbf{H}$. However, the complexity of such a search is exponential.

In order to develop codes for which practical (polynomial complexity) decoding procedures can be developed, it is necessary to place additional restrictions on the linear codes. Important classes of such codes are discussed in Chapter 5.

4.2.4 Estimation of Error Probability of Decoding

As mentioned in Section 4.2.1, it is often most reasonable to decode based on the maximum of the LF, but in many cases such a decoding procedure can make the estimation of the achieved error probability very difficult. On the other hand, many practical coding schemes use decoding in the sphere of radius t (as shown in Figure 4.4), and in this case it is possible to obtain some analytic estimates.

We will assume that the decoder considers the entire received word, and we estimate the error probability of the decoder – the probability that the decoder did not determine the correct word. Such a probability is called the *error probability for a word*.

This probability can be easy to calculate – as, for example, in the case of the BSC. An error takes place when the channel introduces errors into v symbols and the value of v exceeds the radius of the sphere, t. For a BSC with conditional probability p, the probability of the event $v > t$ for n transmitted symbols, considering (4.1), equals

$$P_w = P(v > t) = \sum_{v=t+1}^{n} C_n^v p^v (1 - p)^{n-v} \tag{4.15}$$

Expression (4.15) requires the summation of binomial coefficients, which is not convenient to compute in practice. Often, we are interested in large n and small values of P_w. In this case, there are known estimates. A nice estimate was presented in [8]:

$$P_w < \frac{p(1 - \tau)}{\tau - p} \frac{1}{\sqrt{2\pi n \tau (1 - \tau)}} e^{-n(T_p(\tau) - H(\tau))} \tag{4.16}$$

where $\tau = t/n$, $\tau > p$, and $H(\tau) = -\tau \ln \tau - (1-\tau)\ln(1-\tau)$ is an entropy of the binary ensemble with parameter τ, and $T_p(\tau) = -\tau \ln p - (1-\tau)\ln(1-p)$.

For transfer via an AWGN channel and using hard decoding at the receiver, the value of P_w can be also found by the use of (4.15) and (4.16), as hard decoding makes the AWGN channel look like a BSC with conditional probability defined according to (4.2). In many practical applications, however, one is not required to estimate the probability of making a mistake when decoding a word; one is required to calculate the *error probability for a bit* because even when the output of the decoder is incorrect, most of the bits may be right. The analytical calculation of this value, however, becomes a complicated task.

It is clear that error bits can occur at the output of the decoder only if the word was not decoded correctly. This event occurs in one of two scenarios.

First, the recorded word may be located outside of all the spheres. In this case, the decoder, following the criterion of decoding within a sphere, tries to correct the errors but finds that there is no appropriate codeword. The decoder can report that this word was received, but it has no way to associate it with a codeword. Let us call this event "refusal to decode." Let us consider the situation where the decoder returns the received word without making any changes.

Assuming that the probability of a bit being corrupted is small, the most probable cause of a "refusal to decode" is that $t+1$ errors were received in error (because the first summand will be dominant in (4.15)). In this case, the fraction of the bits in the word in error will be $(t+1)/n$, and the probability that a bit will be incorrectly decoded can be estimated as

$$P_b^{(1)} \approx \left(\frac{t+1}{n}\right) P_w \tag{4.17}$$

The second way bits can be decoded incorrectly is that the recorded word lies in a sphere, but the center of the sphere is not the codeword that was really transmitted. In this case, the decoding procedure produces a valid codeword, but the codeword is incorrect. Here, we say that a *decoding error* occurred. There is no way for the decoder to know that it is not reporting the true codeword.

Again, when the probability of an incorrect bit being received is small, it is much more probable that the recorded word will be in a sphere that is nearest to the sphere within which it belongs. In this case, the decoded codeword will probably be the codeword nearest to the transmitted word. In a code with minimal distance d_0, we can estimate the number of symbols in which the two codewords differ by d_0. Then

$$P_b^{(2)} \approx \left(\frac{d_0}{n}\right) P_w \tag{4.18}$$

We note that because $d_0 = 2t+1$, from (4.17) and (4.18) we get $P_b^{(1)} \approx \frac{1}{2}P_b^{(2)}$, which for small probabilities of error often can be considered as a negligibly small difference.

4.3   Codes with Algebraic Decoding

4.3.1   Cyclic Codes

In sections 4.1 and 4.2, we considered the task of encoding information message having k symbols into codewords with n symbols, with the goal of increasing the reliability of transmission of information across a noisy communication channel. The number of possible words to transmit when each word is of length k is exponentially large, but as we have seen, making use of linear subsets as codes allows one to design effective encoding procedures.

We defined minimal distance decoding as searching for the codeword nearest to the received codeword in some metric. Because there are so many potential codewords, it is generally impractical to solve this problem by brute force methods. It is therefore necessary to look for linear codes with additional structure that will cause decoding to be of polynomial complexity.

In practice, *linear cyclic codes*, that is, linear codes for which a cyclic shift of any codeword is also a codeword, have been found to be quite useful. To consider such codes, let us associate each polynomial $f(x) = f_{n-1}x^{n-1} + \cdots + f_1 x + f_0$ with the vector $f = (f_{n-1}, \ldots, f_0)$. Then, vectors of length n correspond to a polynomial whose degree does not exceed $n-1$.

Let $g(x)$ be a polynomial of degree $r = n-k$ which is a divisor of the polynomial $x^n - 1$. Then, the set of all polynomials of degree not exceeding $n-1$ that are divisible by $g(x)$ are a cyclic code of length n with k information symbols. The polynomial $g(x)$ is called the *generator polynomial* of the cyclic code. Products of arbitrary polynomials $m(x)$ with the codeword $a(x)$ modulo $x^n - 1$ (that is, taking $x^n = 1$) are also codewords. Coding of an information sequence $m(x) = \sum_{i=0}^{k-1} m_i x^i$ using the cyclic code with the generator polynomial $g(x) = g_r x^r + g_{r-1} x^{r-1} + \cdots + g_0$ is performed using the rule

$$m(x) \rightarrow m(x)g(x) \bmod x^n - 1$$

The generating matrix of the cyclic code can be written as

$$
\mathbf{G} = \begin{bmatrix}
g_0 & \cdots & g_r & 0 & \cdots & \cdots & 0 \\
0 & g_0 & \cdots & g_r & 0 & \cdots & 0 \\
\cdots & \cdots & \cdots & \cdots & \cdots & \cdots & \cdots \\
0 & \cdots & \cdots & 0 & g_0 & \cdots & g_r
\end{bmatrix}
\tag{4.19}
$$

We define the *check polynomial* of the code with generator polynomial $g(x)$ as the polynomial

$$h(x) = \frac{x^n - 1}{g(x)} \tag{4.20}$$

The product of any codeword $a(x) = m(x)g(x)$ and the polynomial $h(x)$ is

$$a(x)h(x) = m(x)g(x)h(x) = 0 \bmod x^n - 1 \tag{4.21}$$

Relation (4.21) is one of the most important relations for the codewords. Because $x^n - 1$ is divisible by $h(x)$ with remainder $g(x)$, expression (4.21) is equivalent to the equation

$$a(x) = 0 \bmod g(x)$$

The check matrix of the cyclic code is given by

$$\mathbf{H} = \begin{bmatrix} h_0 & \dots & h_k & 0 & \dots & \dots & 0 \\ 0 & h_0 & \dots & h_k & 0 & \dots & 0 \\ \dots & \dots & \dots & \dots & \dots & \dots & \dots \\ 0 & \dots & \dots & 0 & h_0 & \dots & h_k \end{bmatrix} \tag{4.22}$$

where the coefficients are those of $h(x)$.

In addition to determining an encoding procedure, the generator polynomial of a cyclic code also allows one to estimate the minimal distance of the code it generates. Let us suppose that $g(x)$ is the generator polynomial of a cyclic code G (that is not composed of the entire space), and all the roots of $g(x)$ lie in the field $GF(q)$. Then the roots can be written as powers of η, of a primitive element of the field $GF(q)$. That is, the roots can be written in the form $\eta^{i_1}, \eta^{i_2}, \dots, \eta^{i_r}$. We consider subsequences of consecutive integers from the set $\{i_1, i_2, \dots, i_r\}$ and assume that the longest sequence (among all such sequences) is $m_0, m_0 + 1, \dots, m_0 + d_0 - 2$. It can be shown that the minimal code distance d of the code G (the bound for Bose–Chaudhuri–Hocquenghem [BCH]-codes) can be estimated from below by $d \geq d_0$. This bound for BCH codes allows one to estimate the distance of arbitrary cyclic codes via the analysis of the roots of the code's generator polynomial, giving an estimate of the minimal distance of the cyclic code, which may be lower than the true minimal distance.

4.3.2 BCH Codes

Let us consider the following approach to the construction of the BCH cyclic codes of length $n = q^m - 1$, where q is a prime number, q^m is a prime power, and the code corrects t errors. We choose a primitive polynomial of degree m and construct a finite field $GF(q^m)$. Let η be a primitive element of $GF(q^m)$. Then we find the minimal polynomials $f_j(x)$ for the elements η^j, $j = 1, 2, \dots, 2t$. We take

$$g(x) = lcm\{f_1(x), f_2(x), \dots, f_{2t}(x)\} \tag{4.23}$$

According to [1, 12] the code generated by $g(x)$ is called a *BCH code*, and it has minimal $d \geq d_0 = 2t + 1$. That is, it corrects up to t errors.

In Table 4.1, minimal polynomials are presented for the elements of $GF(2^4)$ and $GF(4^2)$. More tables of minimal polynomials (for other finite fields) can be found in [1, 12].

Table 4.1 Minimal polynomials for the fields $GF(2^4)$ and $GF(4^2)$.

Field $GF(2^4)$		Field $GF(4^2)$	
Field element	Minimal polynomial	Field element	Minimal polynomial
0	0	0	0
1	$x+1$	1	$x+1$
η	x^4+x+1	η	x^2+x+2
η^2	x^4+x+1	η^2	x^2+x+3
η^3	$x^4+x^3+x^2+x+1$	η^3	x^2+3x+1
η^4	x^4+x+1	η^4	x^2+x+2
η^5	x^2+x+1	η^5	$x+2$
η^6	$x^4+x^3+x^2+x+1$	η^6	x^2+2x+1
η^7	x^4+x^3+1	η^7	x^2+2x+2
η^8	x^4+x+1	η^8	x^2+x+3
η^9	$x^4+x^3+x^2+x+1$	η^9	x^2+2x+1
η^{10}	x^2+x+1	η^{10}	$x+3$
η^{11}	x^4+x^3+1	η^{11}	x^2+3x+3
η^{12}	$x^4+x^3+x^2+x+1$	η^{12}	x^2+3x+1
η^{13}	x^4+x^3+1	η^{13}	x^2+2x+2
η^{14}	x^4+x^3+1	η^{14}	x^2+3x+3

The check matrix of the BCH code can be written as

$$
H = \begin{bmatrix}
1 & \eta^{m_0} & \cdots & (\eta^{m_0})^{n-1} \\
1 & \eta^{m_0+1} & \cdots & (\eta^{m_0+1})^{n-1} \\
\cdots & \cdots & \cdots & \cdots \\
1 & \eta^{m_0+d_0-2} & \cdots & (\eta^{m_0+d_0-2})^{n-1}
\end{bmatrix} \tag{4.24}
$$

Note that matrix (4.24) is a *Vandermonde matrix* and is defined over the field $GF(q)$ – an extension of the field over which the code is given. The number of rows in it differs from $r = n - k$. Till now, we got a parity-check matrix over the main field, where the number of rows was strictly equal to r. To pass from matrix (4.24) to the traditional-form matrix it is sufficient to exchange elements η^i in it in the form of vector columns and eliminate any linearly dependent rows from the matrix one obtains in this way.

It is possible to use non-primitive field elements to generate a BCH code. Let γ be an element of $GF(q^m)$ whose multiplicative order equals N, and assume that $m_0 \geq 0$ and $2 \leq d_0 \leq N$ are whole numbers, and, additionally, $f_{m_0}(x), f_{m_0+1}(x), \ldots, f_{m_0+d_0-2}(x)$ are the minimal polynomials for

$\gamma^{m_0}, \gamma^{m_0+1}, \ldots, \gamma^{m_0+d_0-2}$, respectively. Then the cyclic code with the generator polynomial

$$g(x) = lcm\{f_{m_0+1}(x), \ldots, f_{m_0+d_0-2}(x)\} \tag{4.25}$$

is called a *BCH code*, and its parameters are: $n = N$; $k = N - \deg g(x)$, where $\deg(\cdot)$ is degree of the polynomial $(\cdot)$; $d \geq d_0$. If $n = N = q^m - 1$, the BCH code is called *primitive*.

The number of check symbols r of the BCH codes (i.e. degree of the generator polynomial), constructed with the help of the element $\gamma \in GF(q^m)$, can be estimated by the product $r \leq m(d - 1)$. For binary codes, this estimate can be improved as $r \leq mt$. At the same time, we note that correction of the real distance of the cyclic codes and much more precise estimation of the number of check symbols of BCH codes are still open problems.

4.3.3 Reed–Solomon Codes

Reed–Solomon (RS) codes were invented in 1960 and are still one of the most commonly used classes of codes and are, despite their simplicity, or maybe because of it, the basis of a new, deep generalization [1, 12, 40]. Let us start the discussion of RS codes with the following simple remark: For arbitrary linear (n, k)-codes with distance d the following constraint, called the *Singleton bound*, is valid: $n - k \geq d - 1$.

Codes achieving the Singleton bound are optimal in terms of minimal distance, and therefore, they are usually called maximum distance separable (MDS) codes.

In practice, the most important examples of MDS codes are the RS codes, which can be obtained from BCH codes, given over the same field inside of which the roots of the generator polynomial lie, that is, $n = q - 1$. In this case, the generator polynomial for the code that corrects t errors, according to (4.23), equals

$$g(x) = \prod_{i=1}^{2t} (x - \eta^i) \tag{4.26}$$

The check matrix of the RS code in such a case is the Vandermonde matrix of (4.24). Because for the RS codes the Singleton bound is met, $d = n - k + 1$ for these codes. In many practical applications and standards, RS codes are used with the length $n = 255$ over the field $GF(2^8)$ (for example, the $(255, 239)$-code that corrects 8 errors), because symbols from this field can be represented by 8-bit sequences (bytes), which simplify the realization of the coders and decoders.

4.4 Decoding of Cyclic Codes

BCH codes and RS codes are widely used both because of their good distance and speed characteristics and because there are effective decoding procedures that take advantage of the ability to describe the codes using the algebra of polynomials.

Let G be a cyclic (n, k)-code (BCH or RS) over the field $GF(q)$ with distance $d = 2t + 1$, and let $g(x)$ be the generator polynomial of the code. Additionally, assume that $\eta, \eta^2, \ldots, \eta^{2t}$, are the roots of the polynomial $g(x)$. Recall that the check matrix of the code G can be written in the form of (4.24), which (for convenience's sake) we repeat here:

$$\mathbf{H} = \begin{bmatrix} 1 & \eta^{m_0} & \cdots & (\eta^{m_0})^{n-1} \\ 1 & \eta^{m_0+1} & \cdots & (\eta^{m_0+1})^{n-1} \\ \cdots & \cdots & \cdots & \cdots \\ 1 & \eta^{m_0+d_0-2} & \cdots & (\eta^{m_0+d_0-2})^{n-1} \end{bmatrix} \tag{4.27}$$

Let $b(x)$ be a received word – a word that is the sum of the transmitted codeword, $a(x)$, and the error, $e(x)$. That is, $b(x) = a(x) + e(x)$, and the error polynomial is given by

$$e(x) = \varepsilon_0 + \varepsilon_1 x + \cdots + \varepsilon_{n-1} x^{n-1} \tag{4.28}$$

and consists of exactly $v \le t$ nonzero coefficients $\varepsilon_{i_1}, \ldots, \varepsilon_{i_v}$. Multiplying the matrix $\mathbf{H}$ by the vector associated with $b(x)$, we find that the result, called the syndrome, has the components

$$S_j = b(\eta^j), \quad j = 1, \ldots, 2t \tag{4.29}$$

Because $a(\eta^j) = 0$ for any codeword, Eq. (4.29) can be rewritten as

$$S_j = \varepsilon_{i_1}(\eta^j)^{i_1} + \varepsilon_{i_2}(\eta^j)^{i_2} + \cdots + \varepsilon_{i_v}(\eta^j)^{i_v} \tag{4.29a}$$

The error polynomial (4.29a) is fully described by the set of pairs $\{\varepsilon_{i_1}, i_1\}, \ldots, \{\varepsilon_{i_v}, i_v\}$. Denoting ε_{i_s} by Y_s, and η^{i_s} by X_s, we get

$$S_j = X_1^j Y_1 + \ldots + X_v^j Y_v, \quad j = 1, \ldots, 2t \tag{4.30}$$

The values Y_s will be called the *values* of the errors, and X_s will be called the *locators* of the errors. S_j can be calculated directly from the obtained vector, and, therefore, Eq. (4.30) can be considered as a system of $2t$ nonlinear equations relative to $2v$ unknown variables $X_1, X_2, \ldots, X_v$, $Y_1, \ldots, Y_v$, which should enable us to determine the transmitted word from the received word. Historically, the first algorithm for solving (4.30) was the Peterson–Gorenstein–Zincler (PGZ) algorithm [13], which transforms the nonlinear system (4.30) into a linear system. Unfortunately, this algorithm's complexity is $O(n^3)$, and, therefore, it is not used in practical applications because there are now more effective algorithms for decoding cyclic codes. Let us explain what this means.

In order to realize the system of linear equations in the PGZ algorithm, the polynomial $\sigma(x) = 1 + \sigma_1 x + \cdots + \sigma_v x^v$ of minimum degree is introduced, the roots of which, $X_1^{-1}, \ldots, X_v^{-1}$, are measures that are inverse to the error locators, that is,

$$\sigma(x) = (1 - xX_1)(1 - xX_2) \cdots \cdots (1 - xX_v) \tag{4.31}$$

The polynomial $\sigma(x)$ is called the *polynomial of error locators*, and its definition is a central task in all algebraic procedures for decoding cyclic codes. The main difference in these procedures compared to the PGZ algorithm is the use of the so-called *key equation*

$$\frac{\omega(x)}{\sigma(x)} = S(x) \bmod x^r \tag{4.32}$$

instead of the solution of Eq. (4.31) to define not only the polynomial of error locators $\sigma(x)$ but also the polynomial of *error values* $\omega(x)$.

Finally, we present a general scheme for decoding cyclic codes.

(1) *The first step*: *Computation of the syndrome.* Let $\mathbf{b} = \mathbf{a} + \mathbf{e}$ be the received vector, where a is the codeword, $\mathbf{e}$ is the error vector, and $W(\mathbf{e}) \leq t$. The first step in the decoding procedure is to calculate the syndrome $S(x)$ (see (4.32)). Note that this can be done with linear complexity with the help of the Gorner scheme [1, 12, 40], according to which we compute the value of the polynomial $f(x)$ of degree n at the desired point using the following operations:

$$f(x) = \sum_{i=0}^{n} f_i x^i = f_0 + x(f_1 \mid x(f_2 + x(\ldots (f_{n-1} + f_n x)))) \tag{4.33}$$

(2) *The second step*: *Solution of the key equation.* The second phase of decoding is the solution of the key equation – Eq. (4.32). That is, we determine the polynomial of locators $\sigma(x)$ and the polynomial of error values $\omega(x)$. The polynomial $\omega(x)$ will be used in the *fourth step* for error definition.

(3) *The third step*: *Calculation of error locators (Chen procedure [1, 12]).* In this step, we calculate the roots of $\sigma(x)$. Because the roots of this polynomial lie in the limit field $GF(q)$, they can be calculated with the help of rotation searching. Note that for its realization the Gorner scheme [1, 12] can be used.

(4) *The fourth step*: *Calculation of error values (Fordi procedure [1, 12]).* In the last step, the error values can be determined with the help of the polynomial $\omega(x)$, obtained during the procedure carried out in the *second step*:

$$\varepsilon_i = -\frac{\omega(\eta^{-i})}{\sigma'(\eta^{-i})} \tag{4.34}$$

where $\sigma'(x)$ is the formal derivative of the polynomial $\sigma(x)$:

$$\sigma'(x) = \sum_{j=1}^{v} j\sigma_j x^{j-1}$$

and i is the error's position.

A better approach is to make use of the Berlekamp–Massey (BM) iterative algorithm [1, 8] to perform the second and third steps. This means that the polynomial $\sigma(x)$ is defined by making use of the following approximations, beginning with $\sigma^{(0)}(x)$, $\sigma^{(1)}(x)$, and so on, until we reach $\sigma^{(v)}(x) = \sigma(x)$, selecting $\sigma^{(j)}(x)$ as a correction to $\sigma^{(j-1)}(x)$.

For components of the syndrome the following recurrence relation holds:

$$S_j = -\sum_{i=1}^{v} \sigma_i S_{j-i}, \quad j = v+1, \ldots, 2v \tag{4.35}$$

We will declare that the polynomial $\sigma^{(j)}(x)$ *generates* $S_1, \ldots, S_j$, if for all these values the following equation holds:

$$S_j = -\sum_{i=1}^{v^{(j)}} \sigma_i^{(j)} S_{j-i} \tag{4.36}$$

where $v^{(j)}$ is the degree of the polynomial $\sigma^{(j)}(x)$ and $\sigma_i^{(j)}$ are the coefficients of this polynomial. In fields of characteristic 2 (see the second step), this constraint can be rewritten as

$$\Delta^{(j)} = S_j + \sum_{i=1}^{v^{(j)}} \sigma_i^{(j)} S_{j-i} \tag{4.37}$$

If $\Delta^{(j)} = 0$, then $\sigma^{(j)}(x)$ generates $S_1, \ldots, S_j$, and if $\Delta^{(j)} \neq 0$ then $\sigma^{(j)}(x)$ does not generate this sequence of syndromes.

In order to determine $\sigma(x)$ we look for the polynomial of minimal degree, which generates the sequence of the components of the syndrome $S_1, S_2, \ldots,$ S_{2t}. The iterative process of determining $\sigma(x)$ starts with $\sigma^{(0)}(x) = 1$ and consists of the following stages. For a given polynomial $\sigma^{(j-1)}(x)$ generated by $S_1,$ $\ldots, S_{j-1}$, we check to see if $\sigma^{(j-1)}(x)$ is not a generator of S_j (i.e. we check to see whether $\Delta^{(j)} = 0$ holds). If it is a generator, then $\sigma^{(j)}(x)$ should be equal to $\sigma^{(j-1)}(x)$. Conversely, the $\sigma^{(j)}(x)$ are chosen in such a manner that they are the generators of the sequence $S_1, \ldots, S_{j-1}, S_j$. The process continues until the stage at which the corresponding polynomial that generates all components of the syndrome is found.

The algorithm proposed by Berlekamp and Massey has numerous variations (see [1, 8]). In [38, 40], one version of the BM algorithm is presented and described. There, in addition to multiple members of locators, $\sigma(x)$, the algorithm also finds multiple members of values of $\omega(x)$, which can then be used to determine the error values. We do not enter into mathematical descriptions of such algorithms because such complicated mathematical analysis is beyond the scope of this book.

The BM algorithm is probably the best known method for decoding cyclic codes. Its complexity is estimated as $O(n^2)$, but implementation of the algorithm generally makes use of many modifications to decrease the complexity of the practical decoding scheme presented above.

Nevertheless, the BM algorithm is not the only algebraic decoding algorithm. There are, for example, algorithms that are based on the extended Euclidian algorithm and also those that make use of the fast discrete Fourier transform over the finite field [1, 8, 12, 38, 40] in the decoding procedure.

In conclusion, note that as a result of decoding the cyclic code, a codeword is generated at a distance not exceeding t from the recorded (received) word. In Section 4.2, we described this as decoding in a sphere of radius t. In this case, the probability of decoding with errors can be estimated analytically (see Section 4.2.4).

References

1 Blahut, R. (1986). *Theory and Practice of Codes Controlling Errors.* Moscow: Mir in Russian.

2 Blokh, E.L. and Zyablov, V.V. (1976). *Generic Cascade Codes (Algebraic Theory and Complexity in Realization).* Moscow: Svyaz in Russian.

3 Blokh, E.L. and Zyablov, V.V. (1982). *Linear Cascade Codes.* Moscow: Nauka in Russian.

4 Gallager, R. (1974). *Theory of Information and Stable Communication.* Soviet Radio: Moscow (in Russian).

5 Gallager, R. (1966). *Codes with Small Density of Checking on Evenness.* Moscow: Mir in Russian.

6 Zyablov, V.V. and Pinsker, M.S. Estimation of difficulties of errors correction by low-dense Gallager's codes. *Prob. Inf. Transm.* XI (1): 1975, 23–1926. in Russian.

7 Kozlov, A.V., Krouk, E.A., and Ovchinnikov, A.A. (2013). Approach to block-exchanging codes design with low density of checking on evenness. *Izv. Vuzov,* Instrumentation Industry (8): 9–14. in Russian.

8 Kolesnik, V.D. (2009). *Coding in Transmission, Transfer and Secure Information (Algebraic Theory of Block-Codes).* Moscow: Vishaya Shkola in Russian.

9 Kolesnik, V.D. and Mironchikov, E.T. (1968). *Decoding of Cycle Codes.* Moscow: Svyaz' in Russian.

10 Kolesnik, V.D. and Poltirev, G.S. (1982). *Course of Information Theory.* Moscow: Nauka in Russian.

11 Kudryashov, B.D. (2009). *Information Theory.* Piter: SPb in Russian.

12 Mak-Vil'ams, A.D. and Sloen, D.A. (1979). *Codes Theory, Corrected Errors.* Moscow: Svyaz' (in Russian).

13 Piterson, U. and Ueldon, E. (1976). *Codes, Corrected Errors.* Moscow: Mir (in Russian).

14 Forni, D. (1970). *Cascade Codes.* Moscow: Mir in Russian.

15 Shannon, K. (1963). *Works on Information Theory and Cybernetics.* Moscow: Inostrannaya Literatura (in Russian).

16 ITU-TRecommendation G. 975/G709 (2000). *Forward error correction for submarine systems.*

17 ITU-T Recommendation G. 975.1 (2004). *Forward error correction for high bit-rate DWDM submarine systems.*

18 Arabaci, M., Djordjevic, I.B., Saunders, R., and Marcoccia, R.M. (2010). Nonbinary quasi-cyclic LDPC based coded modulation for beyond 100-Gb/s. *IEEE Trans., Photonics Technol. Lett.* 22 (6): 434–436.

19 Bahl, L., Cocke, J., Jelinek, F., and Raviv, J. (1974). Optimal decoding of linear codes for minimizing symbol error rate. *IEEE Trans. Inform. Theory* IT-20 (2): 284–287.

20 Barg, A. and Zemor, G. (2005). Concatenated codes: serial and parallel. *IEEE Trans. Inform. Theory* 51 (5): 1625–1634.

21 Batshon, H.G., Djordjevic, I.B., Xu, L., and Wang, T. (2009). Multidimensional LDPC-coded modulation for beyond 400 Gb/s per wavelength transmission. *IEEE Photonics Technol. Lett.* 21 (16): 1139–1141.

22 Berlekamp, E., McEliece, R., and Van Tilborg, H. (1978). On the inherent intractability of certain coding problems. *IEEE Trans. Inform. Theory* 24 (3): 384–386.

23 Butler, B.K. and Siegel, P.H. (2014). Error floor approximation for LDPC codes in the AWGN channel. *IEEE Trans. Inform. Theory* 60 (12): 7416–7441.

24 Chen, J., Tanner, R.M., Jones, C., and Li, Y. (2005). Improved min-sum decoding algorithms for irregular LDPC codes. *Proc. Int. Symp. Inf. Theory*: 449–453.

25 Cole, C.A., Wilson, S.G., Hall, E.K., and Giallorenzi, T.R. (2006). Analysis and design of moderate length regular LDPC codes with low error floors. *Proc. 40th Annu. Conf. Inf. Sci. Syst.*, pp. 823 and 828.

26 Diao, Q., Huang, Q., Lin, S., and Abdel-Ghaffar, K. (2012). A matrix-theoretic approach for analyzing quasi-cyclic low-density parity-check codes. *IEEE Trans. Inf. Theory* 58 (6), pp. 4030, 4048.

27 Djordjevic, I., Arabaci, M., and Minkov, L.L. (2009). Next generation FEC for high-capacity communication in optical transport networks. *J. Lightwave Technol.* 27 (16): 3518–3530.

28 Djordjevic, I., Sankaranarayanan, S., Chilappagari, S.K., and Vasic, B. (2006). Low-density parity-check codes for 40-gb/s optical transmission systems. *IEEE J. Sel. Top. Quantum Electron.* 12 (4): 555–562.

29 Djordjevic, I. and Vasic, B. (2005). Nonbinary LDPC codes for optical communication systems. *IEEE Photonics Technol. Lett.* 17 (10): 2224–2226.

30 Djordjevic, I., Xu, J., Abdel-Ghaffar, K., and Lin, S. (2003). A class of low-density parity-check codes constructed based on Reed–Solomon codes with two information symbols. *IEEE Photonics Technol. Lett.* 7 (7): 317–319.

31 Djordjevic, I., Xu, L., Wang, T., and Cvijetic, M. (2008). GLDPC codes with Reed–Muller component codes suitable for optical communications. *IEEE Commun. Lett.* 12 (9): 684–686.

32 Forestieri, E. (2005). *Optical Communication Theory and Techniques*. Germany: Springer.

33 Forney, G.D., Richardson, T.J., Urbanke, R.L., and Chung, S.-Y. (2001). On the design of low density parity-check codes within 0.0045 dB of the Shannon limit. *IEEE Commun. Lett.* 5 (2): 214–217.

34 Fossorier, M., Mihaljevic, M., and Imai, H. (1999). Reduced complexity iterative decoding of low density parity-check codes based on belief propagation. *IEEE Trans. Commun.* 47 (5): 379–386.

35 Guruswami, V. and Sudan, M. (1999). Improved decoding of Reed–Solomon and algebraic geometric codes. *IEEE Trans. Inform. Theory* IT-45 (6): 1755–1764.

36 Hirotomo, M. and Morii, M. (2010). Detailed evaluation of error floors of LDPC codes using the probabilistic algorithm. *Proc. Int. Symp. Inf. Theory Its Appl. (ISITA)*: 513–518.

37 Gho, G.-H. and Kahn, J.M. (2012). Rate-adaptive modulation and low-density parity-check coding for optical fiber transmission systems. *IEEE/OSA J. Opt. Commun. Networking* 4 (10): 760–768.

38 Kabatiansky, G., Krouk, F., and Semenov, S. (2005). *Error Correcting Coding and Security for Data Networks: Analysis of the Super Channel Concept*. Wiley.

39 Kou, Y., Lin, S., and Fossorier, M.P.C. (2001). Low-density parity-check codes based on finite geometries: A rediscovery and new results. *IEEE Trans. Inf. Theory* 47 (7): 512–518.

40 Krouk, E. and Semenov, S. (eds.) (2011). *Modulation and Coding Techniques in Wireless Communications*. New Jersey: Wiley.

41 Li, J., Liu, K., Lin, S., and Abdel-Ghaffar, K. (2015). A matrix-theoretic approach to the construction of non-binary quasi-cyclic LDPC codes. *IEEE Trans. Commun.* 63 (4): 1057–1068.

42 Lin, S. and Ryan, W. (2009). *Channel Codes: Classical and Modern*. Cambridge: Cambridge University Press.

5

Coding in Optical Communication Channels

5.1 Peculiarities of Cyclic Codes in Communication Systems

In Chapter 4, cyclic codes were introduced as a classical result of coding theory. The relation between these codes and the algebra of polynomials allows us to obtain polynomial-based procedures for decoding cyclic codes. At the same time, these codes each have their idiosyncrasies, which should be taken into account when selecting codes and code parameters for specific applications. The topic of this chapter makes use of the ideas discussed in [1–54].

As shown in Section 4.1.1, classic algebraic decoders for cyclic codes work in the same finite field over which the code was given. Such decoders can be used, for example, for binary codes for the binary symmetric channel (BSC) – see Section 4.1.2 – but it is more challenging to use them when the signal is being transmitted over a semicontinuous channel, such as the additive white Gaussian noise (AWGN) channel (see Section 4.1.3). This is because the decoders are designed to process symbols from a finite field and make use of the algebraic properties of the code being used but are not designed to assign symbols on a probabilistic basis to a received waveform or analog signal.

Constructive distance: The Bose–Chaudhuri–Hocquenghem (BCH) and Reed–Solomon (RS) codes discussed in Sections 4.3.2 and 4.3.3, respectively, can be considered codes with the parameters n, k, and d_0, which are able to correct arbitrary combinations of $t = \lfloor (d_0 - 1)/2 \rfloor$ errors; in this case, the real minimal distance d of the cyclic codes is generally not known and is estimated as $d \geq d_0$. The parameter d_0 is called the constructive minimum distance, and the codes are certainly capable of correcting errors whose weight does not exceed half the design distance.

Decoding within a sphere: The constructive distance provides a guaranteed lower bound for the code distance and influences the decoder, because the decoders described in Section 4.1 can guarantee the correction of not more than t errors. In other words, the coders guarantee decoding when

Fiber Optic and Atmospheric Optical Communication, First Edition.
Nathan Blaunstein, Shlomo Engelberg, Evgenii Krouk, and Mikhail Sergeev.
© 2020 John Wiley & Sons, Inc. Published 2020 by John Wiley & Sons, Inc.

the received word is located in one of the spheres but do not work when the received word is not contained in any decoding sphere. There are known coding procedures that will decode beyond the constructive distance (for example, the Guruswami–Sudan (GS) algorithm [35, 38]), but they are more computation intensive.

Block decoding: As mentioned in Section 4.1, an algebraic decoder decodes relative to spheres of radius t and considers the whole block. Thus, when estimating the error probability it is natural to estimate the error probability over the transmitted block. To decrease this probability, one must construct codes with large minimal (or constructive) distances. In practice, many transmission systems are described by the probability of error in a transmitted bit. Minimizing this value can be achieved efficiently by the use of decoders that work with symbols, which is not the traditional method used for cyclic codes. Such symbol-to-symbol decoding can be achieved by presentation of the code in the form of the so-called trellises and decoding with their help [19]. In the case of cyclic codes, this procedure is exponentially complicated and is only practical for codes with comparatively small parameter values.

In Section 5.2, we discuss one class of codes whose characteristics differ from those of the codes mentioned above.

5.2 Codes with Low Density of Parity Checks

5.2.1 Basic Definitions

In Section 4.2.2 cyclic codes were considered, and in Section 4.2.3 their principal characteristics were described. Over the preceding two decades, the development of coding theory has been characterized by dealing with semicontinuous communication channels. Construction of the decoders in these cases is, as a rule, based on iterative intra-symbol procedures, and the error probability is not defined by the minimal distance of the corresponding codes, but, more importantly, depends on the characteristics of the decoder.

In such a situation, it is difficult to determine the characteristics of the code construction and to develop an analytic estimate of the error probability. In order to resolve both problems, one often makes use of pseudo-random codes, often based upon heuristic-based construction techniques, and estimates the error probabilities via intensive computer modeling. Making use of turbo-codes or low-density parity check (LDPC) codes, coding schemes that are much more effective than classical cyclic codes with "hard" block-to-block decoding can be achieved.

As a rule, turbo-codes are most effective in low-rate communication systems, and LDPC codes are most effective for high rates and are widely used in high-rate communication systems. Moreover, LDPC codes are particularly effective when used for transmission along an optical channel. The following paragraph discusses LDPC codes.

LDPC codes were proposed comparatively recently by R. Gallager [5] and later in [6, 42–44, 46]. Despite the fact that for a long period LDPC codes were practically eliminated from consideration, over the last few years there has been a large amount of research in this area. This fact relates to the following: Although low-density codes have small minimal distance, LDPC codes make it possible to correct large numbers of errors with algorithms that are not too complex. It was shown that as block length increases, some LDPC codes can outperform turbo-codes, and when used in an AWGN channel, the codes can enable the transfer of information at rates that approach the capacity of the AWGN channel [33].

Many of the constructions proposed for LDPC codes are cyclic or quasi-cyclic and allow both fast decoding and the use of especially effective coding procedures. Even for LDPC codes that are not cyclic, effective procedures were proposed [47].

Traditionally, LDPC codes are described by their parity-check matrix, **H**. For such codes, the parity matrix has the low-density property: Its rows and columns have small numbers of non-zero terms (compared to the matrix's dimension). As linear block codes are often described as the null space of the parity-check matrix, LDPC codes are often given with the help of the incidence graph of the matrix **H** (e.g. the Tanner graph [51]). Such a graph is a 2D graph, the nodes of which are divided into two sets: n symbol nodes, corresponding to columns of the matrix, and r check nodes, corresponding to the rows of the parity-check matrix. The edges connecting the nodes of the graph correspond to non-zero entries in the matrix **H**. An example of such a graph is given in Figure 5.1.

Figure 5.1 An example of a Tanner graph.

Symbol nodes   Check nodes

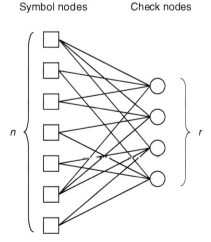

The iterative decoding algorithm for data encoded using LDPC codes finds each symbol separately. Even if the channel is very noisy, the number of errors after the decoding may be small. At the same time, parameters other than the minimal distance may also affect the performance of the algorithm.

Let g_0 be the minimal length of a cycle in the Tanner graph related to the LDPC code. (This value is called the girth.) In [5], it is shown that the iterative decoder provides an exponential decrease in the decoding error probability, provided that the values computed by the decoder are statistically independent, which is achieved for sufficiently large values of g_0.

Apart from the minimal distance d_0 and the length of the minimal cycle g_0, the important characteristics of LDPC codes are the distributions of the non-zero elements in the rows and columns of the parity-check matrix. LDPC codes for which the rows and columns contain equal numbers of units are called *regular codes*, whereas codes that do not contain equal numbers of units are called *irregular codes*.

Let us define a regular (n, γ, ρ)-code as a linear code of length n, for which each column of the parity-check matrix consists of γ non-zero elements, and each row of the parity-check matrix has ρ non-zero elements. Irregular codes often are given by making use of their weight distribution functions, $\lambda(x)$ and $\rho(x)$, where $\lambda(x) = \sum_{i=2}^{d_v} \lambda_i x^{i-1}$ and $\rho(x) = \sum_{i=2}^{d_c} \rho_i x^{i-1}$, where λ_i is the ratio of the columns of the parity-check matrix that have weight i to the total number of columns, and ρ_i is the ratio of the rows in $\mathbf{H}$ of weight i to the total number of rows, and d_v and d_c are the maximum weights of the columns and the rows, respectively.

As a rule, probabilistic methods are used for the construction of good irregular codes, and the asymptotic behavior of such codes is analyzed. On the other hand, regular codes are constructed by making use of (mathematical) objects with known properties and can be analyzed by considering the properties of these objects. Careful selection of the distribution of the weights of an irregular code may be beneficial – especially when the signal-to-noise ratio (SNR) is low (in AWGN channels), when the quality of the code is defined by its **average** characteristics. As the SNR increases, the probability of a decoding error becomes more dependent on the distance characteristics of the code. Here, regular codes that are easier to analyze can be selected to maximize the minimum distance and to have the best spectral properties.

The definition of LDPC codes as linear codes having a parity-check matrix with a small number of non-zero elements does not provide a concrete method of constructing parity-check matrices. Moreover, the construction of LDPC codes often uses probabilistic methods. There are results for LDPC codes, and they are based on the analysis of the set of codes and on the estimation of the codes' characteristics via averaging over the set. A set of LDPC codes is given, first of all, by the weights of the rows and the columns (γ, ρ) of the parity-check matrix, and the code length n is generally the parameter estimated asymptotically for the given weights.

In [5], Gallager showed that for the set of (n, γ, ρ)-codes there exists a parameter δ_{jk} such that as the code length n grows, most codes in the set have minimal distance $n\delta_{jk}$, where δ_{jk} does not depend on n, and therefore, the minimal distance of most codes in the set increases linearly with n.

Gallager proposed algorithms to decode LDPC codes that can work both when the output of the channel is discrete and when it is continuous. Pinsker and Zyablov [6] have shown that among LDPC codes there exist codes for which the decoding algorithm allows for the correction of a number of errors that is proportional to δn and that has complexity of order $n \log n$.

Information theory puts a limit on the rate at which information can be transferred across a communication channel – on a channel's capacity, and that limit is related to a channel parameter. As the parameter grows, the capacity of the channel decreases. For the BSC channel, it can be taken to be the conditional probability, and for an AWGN channel it can be taken to be the noise power per hertz.

In [46, 48], a procedure that allows one to calculate the limit for some channels and to define those weight distributions (γ, ρ) (or $(\lambda(x), \rho(x))$ for irregular codes) that maximize a parameter of the limit is introduced. This allows one to find weight distributions that lead to a high-quality code – mostly for transmission over the channel when the parameter is close to its limiting value (for example, at low SNR).

This procedure is called density evolution, because it calculates the evolution of the probability density of the number of error messages with the number of decoding iterations.

This analysis is asymptotic and does not give a method of constructing codes with a given weight distribution. Frequently, the distribution of weights obtained using this technique are used by different constructions, initially irregular and regular, and then the weights are optimized further.

Finally, an important effect associated with LDPC codes is the "error floor" effect: The rate at which the probability of error decreases slows down as the SNR in the channel increases. This effect is most obvious for irregular designs due to the optimization of weight distributions, which leads to a sharp decrease in the probability of error in the lower SNR region, but this leads to less rapid improvements for large SNR values. Usually, the existence of the saturation effect is due to the fact that the minimum distance of the LDPC codes is small. In this case, the relatively rare event of uncorrected low-weight errors does not greatly affect decoding for small SNRs, but they become the dominant factor when the SNR is large.

This effect is also related to the performance of the iterative decoder, in which the reliability of the decoded value of each symbol depends on that of a small number of other symbols. This problem is usually described by stating that there is a *trapping set* [46, 48]. Building good codes that do not have an "error floor" remains a problem that is still far from having a good, general solution.

It is common practice to search for such codes using mathematical modeling and various heuristic approaches [23, 25, 36, 50].

5.2.2 Decoding of LDPC Codes

Gallager introduced decoding procedures for LDPC codes both for discrete and semicontinuous channels [5]. These procedures were meant to correct errors in channels without memory, where the distortion of one position in the transmitted codeword does not depend on the distortions occurring in other positions. Decoding techniques have been modified many times with the aim of simplifying their realization and decreasing their associated error probabilities. All of these decoding techniques make it possible to iteratively decode the received word from "symbol to symbol" [42].

Initially, we consider the decoder for a discrete channel, and we take the BSC channel as an example. The idea of the "hard" decoder is as follows. Since the verification matrix of the LDPC code contains few units, each code symbol is contained in a small number of checks. Therefore, the erroneous symbol enters the verification relationships of a small number of code symbols and with high probability does not significantly affect the checks of the next decoded symbol. Therefore, if there are not very many mistakes, most of the checks in which the decoded symbol enters are not distorted. Then, the correct decision about the received symbol can be made based on the majority of the checks. Of course, although this technique often works and the correct value of the symbol is chosen, in some cases this technique will introduce an error. It has been proved that by repeating such a majority procedure it is possible, as a rule, to reduce the number of errors. This leads to the following decoding procedure – which is called *bit inversion* (or "bit-flipping").

Initially, all parity checks given by the matrix **H** are calculated. (That is, the syndrome's components are calculated.) Then, the bits of the received word that appear in the calculation of many parity checks that were not initially satisfied are inverted (flipped), and the procedure is repeated as long as a new codeword has not been received and the maximum number of iterations has not been exceeded.

In (soft) decoder design for semicontinuous channels, our certainty about the value taken by the received signal is part of the decoder's input. Our certainty can be represented by the LLR (log likelihood ratio) introduced in (4.6) (see Chapter 4):

$$\text{LLR}(x \mid y) = \log \frac{\Pr(x = +1 \mid y)}{\Pr(x = -1 \mid y)} \tag{5.1}$$

For a bipolar Gaussian channel with variance σ^2, the LLR for the transmitted symbol $x \in \{\pm 1\}$ given that the received symbol was y is $\text{LLR}(x \mid y) = 2y/\sigma^2$ [42].

The **belief propagation** algorithm, which is the standard algorithm for decoding data encoded using LDPC codes, was proposed by Gallager [5]. It can be described by the use of a graphic presentation of the parity-check matrix in the form of the Tanner graph [42], as follows. To all n symbol nodes of the Tanner graph one associates the LLR of the corresponding symbols of the received word. Then the decoder performs iterations, each of which consists of two phases. In the first, the "vertical" phase, each symbol node j, $j = 1, \dots n$ transmits the value $L_{j \to i}$ to each neighboring check node i. The value depends on all the values obtained by the symbol node j from all its neighboring check nodes. The "horizontal" phase works similarly, but in this phase, values of $L_{i \to j}$ are calculated and transmitted from the check to the symbol nodes.

After each iteration, the algorithm makes a decision regarding each code symbol according to the sign of the current value of the node associated with the symbol. If the obtained vector is a valid codeword or if the maximal number of iterations has been reached, the algorithm terminates.

Let us denote the set of check nodes that are neighbors of the symbol node j by $N(j)$; then we denote all neighboring nodes except for the node i, as $N(j)\backslash\{i\}$. Similarly, $N(i)\backslash\{j\}$ is the set of all symbol nodes that are neighbors of check node i, except for the symbol node j.

Decoding LDPC codes can be performed in a reasonably efficient way using belief propagation. There are other algorithms for decoding LDPC codes that are still more efficient and only cause a relatively insignificant deterioration in the error probability [24, 34, 42, 54].

5.2.3 Construction of Irregular LDPC Codes

As mentioned in Section 5.2.1, the density evolution technique allows us to find the distribution of the weights of the rows and columns of the parity-check matrix of irregular LDPC codes. Using such codes increases the work of the iterative decoder – mostly in regions of low SNR.

On the other hand, the work of the decoder is affected by the girth of the Tanner graph (i.e. the minimum length, g_0). The problem arises of constructing a test matrix that corresponds to a given weight distributions $\lambda(x)$ and $\rho(x)$ and at the same time has the greatest possible girth.

In [53, 54], an empirical procedure to solve this problem, a procedure called PEG, "progressive-edge-growth," is proposed. The PEG construction is based on the weight distribution that is taken before computations of symbol and check nodes of the Tanner graph, but exactly the same algorithm can use any arbitrary distribution, including the regular one.

The algorithm with given $\lambda(x)$ and $\rho(x)$ makes use of an iterative procedure that adds edges "step-by-step" to the existing graph. This is done with the aim of maximizing the so-called local minimal cycle for a given node – the length of the minimal cycle that includes this node.

The PEG procedure was introduced and explained in detail in [38, 40]. We do not enter into a deep mathematical description of this procedure and will describe it purely phenomenologically, referring the reader to the excellent works [38, 40, 54]. Let the Tanner graph consist of n symbol nodes v_i, $i = 1, \ldots, n$, and r check nodes c_j, $j = 1, \ldots, r$. Further, we introduce d_{v_i} and d_{c_j}, the degrees of the symbol node v_i and the check node c_j, respectively. We define the degree of a node as the number of edges leaving the node. These values are determined by the distributions $\lambda(x)$ and $\rho(x)$, by a set of edges leaving a symbol node v_i. Let $N_{v_i}^{\ell}$ be the set of characters, which can be reached from the character v_i in ℓ steps. The set $\overline{N}_{v_i}^{\ell}$ is the complement of $N_{v_i}^{\ell}$; it is the set of symbols that satisfies $N_{v_i}^{\ell} \cup \overline{N}_{v_i}^{\ell} = V_c$.

In [54], estimates of the length of the minimal cycle d_0 for codes constructed using this procedure were obtained. Let d_v and d_c be the maximal weights of the symbol and check nodes of the Tanner graph, respectively. Then, the length of the minimal cycle of this graph is bounded from below: $g_0 \geq 2(\lfloor x \rfloor + 2)$, where

$$x = \frac{\log \left(rd_c - \frac{rd_c}{d_v} - r + 1 \right)}{\log((d_v - 1)(d_c - 1))} - 1 \tag{5.2}$$

We now consider Tanner graphs with regular symbol nodes having the same constant degree d_v, and the graph – with the length of the minimum cycle that equals g_0. Then, the minimal distance of the code, given by that graph, can be bounded from below by

$$d_0 \geq \begin{cases} 1 + \dfrac{d_v((d_v - 1)^{\lfloor (g_0 - 2)/4 \rfloor} - 1)}{d_v - 2} & g_0/2 \text{ odd} \\[4mm] 1 + \dfrac{d_v((d_v - 1)^{\lfloor (g_0 - 2)/4 \rfloor} - 1)}{d_v - 2} + (d_v - 1)^{\lfloor (g_0 - 2)/4 \rfloor} & g_0/2 \text{ even} \end{cases} \tag{5.3}$$

Modeling shows that the PEG construction allows us to design irregular codes that lead to a very fast decrease in the error probability and that are often better than regular codes. As was shown earlier, these codes suffer from the saturation effect – from the "error floor" effect. Because the procedure is empirical and pseudo-random and, additionally, depends on input weighted distributions, the quality of the codes obtained using this technique is harder to analyze than those obtained using regular LDPC codes. The code's performance is, therefore, often characterized using intensive modeling.

5.2.4 Construction of Regular LDPC Codes

Regular LDPC codes are those for which all the rows of the parity-check matrix have the same weight and all of the columns have the same weight (which need not be the same as the weight of the rows). (See the definitions in Section 5.2.1.) Alternatively, such codes can be characterized as the codes for which all of the symbol nodes of the associated Tanner graph have the same degree as do all the check nodes (though, once again, the common weight of the symbol nodes does not have to be the same as the common weight of the check nodes).

Although this definition does not limit the methods of codes construction, in practice constructing an LDPC code usually requires the use of techniques that guarantee the necessary regularity.

Construction methods that build up LDPC codes from objects with known properties make use of these properties to obtain codes with the given parameters – such as the code length or the code rate. On the basis of these parameters and the properties of the objects, the minimal distance and the length of the minimal cycle of code are estimated.

By the nature of the density evolution procedure, an irregular weight distribution allows us to obtain much lower values of the threshold (see Chapter 4) than can be obtained by using regular distributions. For this reason, we often make use of an approach that makes use of the regular code construction, which allows us to estimate the code parameter (such as the minimal distance or the length of the minimal cycle). Then the regular matrix is transformed into an irregular one, whose weight distribution is close to the desired irregular weight distribution.

Many of the questions surrounding the construction of regular LDPC codes have been resolved [26, 38–42, 52–54]. Most of the constructions can be described with the help of the approach first proposed by Gallager in his pioneering work [5] about the construction of the parity-check matrix from blocks, called *submatrices:*

$$H = \begin{bmatrix} H_{1,1} & H_{1,2} & \cdots & H_{1,\rho} \\ H_{2,1} & H_{2,2} & \cdots & H_{2,\rho} \\ \cdots & \cdots & \cdots & \cdots \\ H_{\gamma,1} & H_{\gamma,2} & \cdots & H_{\gamma,\rho} \end{bmatrix} \tag{5.4}$$

We define the parity-check matrix as

$$H = \begin{bmatrix} C^{i_{11}} & C^{i_{12}} & \cdots & C^{i_{1\rho}} \\ C^{i_{21}} & C^{i_{22}} & \cdots & C^{i_{2\rho}} \\ \cdots & \cdots & \cdots & \cdots \\ C^{i_{\gamma 1}} & C^{i_{\gamma 2}} & \cdots & C^{i_{\gamma\rho}} \end{bmatrix} \tag{5.5}$$

Here $\mathbf{C}$ is the $(m \times m)$-cyclic permutation matrix:

$$\mathbf{C} = \begin{bmatrix} 0 & 0 & 0 & \cdots & 0 & 1 \\ 1 & 0 & 0 & \cdots & 0 & 0 \\ 0 & 1 & 0 & \cdots & 0 & 0 \\ \cdots & \cdots & \cdots & \cdots & \cdots & \cdots \\ 0 & 0 & 0 & \cdots & 1 & 0 \end{bmatrix} \tag{5.6}$$

and $\rho \leq m$. The codes associated with the parity-check matrices of this form are regular LDPC codes of length $n = m\rho$, of column weight γ, and row weight ρ.

In [40] a different structure for the parity matrix, H_W, was proposed. The authors propose constructing the parity matrix like a Vandermond matrix where the elements of the matrix – the submatrices – are powers of the permutation matrix:

$$\mathbf{H}_W = \begin{bmatrix} \mathbf{I}_m & \mathbf{I}_m & \cdots & \mathbf{I}_m \\ \mathbf{I}_m & \mathbf{C} & \cdots & \mathbf{C}^{\rho-1} \\ \cdots & \cdots & \cdots & \cdots \\ \mathbf{I}_m & \mathbf{C}^{s-1} & \cdots & \mathbf{C}^{(\gamma-1)(\rho-1)} \end{bmatrix} \tag{5.7}$$

Such codes have a minimal distance that satisfies $\gamma + 1 \leq d_0 \leq 2m$.

The block exchange construction can be further modified by allowing the block consisting of the matrix $\mathbf{H}_{i,j}$ to be the zero matrix. First, this makes it possible to obtain irregular LDPC codes and to optimize the weight distributions of the rows and columns. Additionally, such a procedure can lead to a decrease in the number of codewords of small weight and, therefore, even if does not lead to an increase in the minimal distance, it may improve the code's spectrum and decrease the error probability.

Selection of locations for the zero blocks in the matrix is a separate task and depends on a view of the parity-check matrix. Some techniques that maintain the regular matrix structure are shown in Figure 5.2. Currently, LDPC codes are seen both as a powerful error correction technique used in many standards and as a fertile research topic with many potential applications.

These codes are used in many places to transfer information through optical communication channels [18, 21, 27–29, 31, 37].

5.3 Methods of Combining Codes

Algebraic codes, such as the BCH and RS codes that were introduced in Section 4.3, are products of classical coding theory. As their length increases, one finds that the minimum distance of the codes does not increase as quickly as the code length, and the complexity of the decoder also increases. These

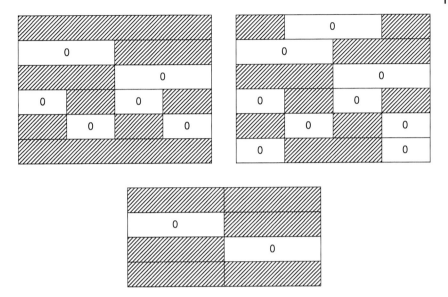

Figure 5.2 Techniques for placing the zero blocks in the parity-check matrix.

codes are oriented toward correcting independent discrete errors, which correspond, for example, to errors in a BSC. In practical channels, errors often suffer from bursts – contiguous regions in which all the bits in the region are corrupted. In order to deal with this problem, one can make use of interleaving – of reordering the bits in a transmission so that bits that are adjacent to one another are not transmitted after one another. Although interleaving helps deal with bursts, we are still left with the problem of constructing long codes (of several thousand characters) having relatively simple decoders.

One of the methods for dealing with this problem is to combine several smaller codes. This approach was pioneered by D. Forni [14] and was called *concatenating codes.* A scheme for concatenating codes makes use of a sequence of connected coders on the transmitting side and a sequence of cascaded decoders on the receiving side, as shown in Figure 5.3.

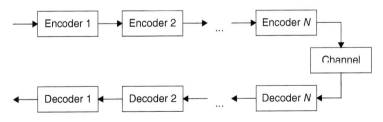

Figure 5.3 Cascading codes.

In the classical approach used in many communication systems two codes are concatenated. The "outer" code is an RS code, and the "inner" code is a BCH code.

Such a scheme also allows one to combine the ability of the binary BCH code to correct independent errors and that of the RS code on the field $GF(2^m)$ to correct the bursts (groups) of binary errors of length m. That is, if the outer code is taken to be the (n_1, k_1, d_1)-code with rate $R_1 = k_1/n_1$, and the inner code is taken to be the (n_2, n_1, d_2)-code with rate $R_2 = n_1/n_2$, the resulting code will be an (n_2, k_1, d_3)-code with rate $R_3 = k_1/n_2 = R_1 R_2$ and minimal distance $d_1 d_2$.

The iterative scheme (also called the code product) is another method of combining codes. In this scheme, the union of the codes is brought about by forming a two-dimensional matrix in which the columns are the codewords of the first code, but the rows are the codewords of the second code. See Figure 5.4.

The product of two codes, an (n_1, k_1, d_1)-code with rate R_1 and an (n_2, k_2, d_2)-code with rate R_2, leads to an $(n_1 n_2, k_1 k_2)$-code with minimal distance $d_3 = d_1 d_2$ and rate $R_3 = R_1 R_2$. Moreover, the 2D nature of the codeword makes it possible to adapt to the channel. Such a code can not only correct all the errors that the code's distance allows, but it can also correct certain combinations of errors that can be described geometrically with the help of a matrix. Thus, a burst that affects a row of the codeword matrix (or even several rows) and contains errors in many symbols can be corrected because the number of errors in the columns remains low. This property is important in channels that can be described as two dimensional – for example, wireless channels for multifrequency transmission that can be thought of as having time and frequency components.

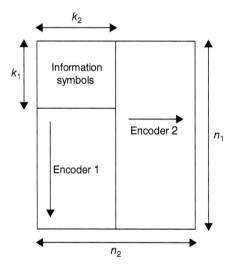

Figure 5.4 Iterative scheme of coding.

It should be noted that the description of concatenated and iterative schemes given above is somewhat simplified. Practical coding schemes that make use of the principles described above can involve a variety of possible modifications to the schemes described above (see [2, 3, 20]).

Further improvement to the reliability of the iterative coding scheme can be achieved by using soft decoding in the binary-input continuous-output communication channel. The component codes in the iterative (or concatenated) schemes usually work with "soft" input and produce "soft" output. Using soft decoding allows the components to pass "soft" information between the different stages of decoding of the different component codes. This idea is of central importance for the production of turbo product codes (TPC), proposed in [45], and it has found many practical applications.

5.4 Coding in Optical Channels

In this section, we consider how the coding techniques described above can be applied to optical communications. Historically, the use of codes for transmission along optical channels can be divided into several stages or "generations" [53, 54].

1. The first generation of error correcting codes used in optical communication systems is described in ITU-T G.709 and G.975 [16]. The encoding schemes described in these standards were based on the classical cyclic codes, BCH and RS codes discussed in Section 4.3, and used hard decoding.
2. The **second** generation of coding schemes is described in the ITU-T G.975.1 standard [17] and makes use of a combination of BCH and RS codes with hard decoding approaches, as discussed in Section 4.3.
3. The **third** generation of coding schemes use the possibility of soft decoding for the LDPC codes in Section 5.2. Selection of LDPC codes instead of, say, turbo codes, was done to enable extremely high-rate data transmission along optical channels, for which LDPC codes are more effective. However, iterative schemes with soft decoding (TPC) as described in Section 5.3 can also be used.
4. Finally, the **fourth** generation is oriented toward extremely high data transmission rates. When considering such systems, the AWGN model is inadequate. It is necessary to take into account other, more specific, characteristics of the channel. One of the issues that must be considered is inter-symbol interference (ISI). In order to overcome ISI, equalization is used. The task of the equalizer is to process the received signal in such a manner that the output noise of the equalizer is nearly Gaussian. This allows us to use standard methods of coding for an AWGN channel. Fourth generation coding schemes for optical communication channels deal primarily with turbo-equalization and the use of LDPC codes.

In coding theory, when it is impossible to obtain analytical estimates of the error probability, computer modeling is usually used. In this case, the ability of the code to correct errors is estimated by the frequency of errors in the information bits or the *bit error rate* (BER) for a given SNR. In practical schemes for the transmission of data in optical channels, one usually estimates the "profit" one makes by making use of coding techniques. This is known as the process or *net coding gain* (NCG), defined as [32]

$$NCG = 20\log_{10}(\mathrm{erfc}^{-1}(2BER_{ref})) - 20\log_{10}(\mathrm{erfc}^{-1}(2BER_{in})) + 10\log_{10}R \text{ (dB)}$$

$$(5.8)$$

where BER_{in} denotes the BER at the input of the decoder, BER_{ref} denotes the reference BER, $\mathrm{erfc}(\cdot)$ is the complementary error function

$$\mathrm{erfc}(x) = \frac{2}{\sqrt{\pi}} \int_x^{+\infty} e^{-z^2} dz \qquad (5.9)$$

and R is the code rate. Today, the standard reference BER for transmission of data along an optical channel is on the order of 10^{-15}.

The current principle standards that describe coding rules for optical systems are the ITU-T G.975 [16] and G.975.1 [17] standards. Additionally, over the course of the preceding few years a set of publications that propose different coding schemes for high-speed optical communication channels (with more than 100 Gbps) have appeared [52].

The ITU-T G.975 standard calls for the use of the classic RS-code (255,239) over the field $GF(2^8)$. In Chapter 4, we considered an estimate of the error probability for a word and for a bit for decoding in the sphere of radius t. In the case of an RS code, this estimate can be corrected because the minimal distance and the weight spectra for these codes are well known.

For example, according to the ITU-T G.975 standard, the error probability over one bit for the code RS (255, 239) with length $t = 8$ can be estimated as

$$P_b = 1 - (1 - P_s)^{1/8} \qquad (5.10)$$

Here P_s is the error probability over a coding symbol consisting of 8 bits, estimated as

$$P_s = \sum_{i=9}^{255} \frac{i}{255} C_{255}^i p_s^i (1 - p_s)^{255-i} \qquad (5.11)$$

where p_s is the error probability for the 8th bit of a symbol in the channel.

The ITU-T G.975.1 standard is the successor of the G.975 standard and proposes nine coding schemes. The parameters of these schemes are shown in Table 5.1, where the first row corresponds to the code G.975. As can be seen, six of the nine coding schemes are iterative or concatenated schemes (the symbol "+" in table denotes the continuous "step-by-step" coding, shown in Figure 5.3), which can use different methods of combining. Most schemes have

Table 5.1 Coding schemes from standards G.975 and G.975.1.

No	Cascade/iterative	Parameters of scheme	Redundancy (%)
1	No	RS (255,239)	6.69
2	Yes	RS (255,239) + convolutional code	24.48
3	Yes	BCH (3860,3824) + BCH (2040,1930)	6.69
4	Yes	RS (1023,1007) + BCH (2040,1952)	6.69
5	Yes (SD)	RS (1901,1855) + iterative code	6.69
6	No	LDPC (32640,30592)	6.69
7	Yes	Cascade BCH	7, 11, 25
8	No	RS (2720,2550)	6.67
9	Yes	Iterative BCH (1020,988) × 2	6.69

Table 5.2 Error correction schemes used in the G.975 and G.975.1 standards.

No	Parameters of scheme	Redundancy (%)	NCG (dB)	Input BER	The target BER
1	RS (255,239)	6.69	5.6	1.8e−4	1e−12
2	RS (255,239) + equalizing code	24.48	8.88	5.2e−3	1e−15
3	BCH (3860,3824) + BCH (2040,1930)	6.69	8.99	3.15e−3	1e−15
4	RS (1023,1007) + BCH (2040,1952)	6.69	8.67	2.17e−3	1e−15
5	RS (1901,1855) + iterative code	6.69	8.5/9.4	1.9/4.5e−3	1e−15
6	LDPC (32640,30592)	6.69	8.02	1.12e−3	1e−15
7	Cascade BCH	7	8.09	1.3e−3	1e−15
		11	9.19	4.44e−3	
		25	10.06	1.3e−2	
8	RS (2720,2550)	6.67	8	1.1e−3	1e−15
9	Iterative BCH (1020,988) × 2	6.69	8.67	3.5e−3	2.1e−14

a redundancy of about 7%. Only the fifth scheme assumes that soft decoding is used (and it is denoted by the abbreviation "SD" in the table).

In Table 5.2, estimates of the noise stability of these schemes in the terminology of NCG (see (5.8)) are shown. Practically all the schemes have a reference BER of about 10^{-15}.

Table 5.3 Coding schemes for transmission above 100 Gbps with approximately 7% redundancy.

Title	Type	HD/SD	Redundancy (%)	NCG (dB)	BER
Swizzle	LDPC	HD	6.7	9.45	1e−15
Staircase	Convolutional DPC	HD	7	9.41	1e−15
SP-BCH	Iterative BCH	HD	7	9.4	1e−15
2-iterative concatenade BCH	Cascade BCH	HD	6.81	8.91	1e−15
UEP-BCH	Iterative BCH	HD	7	9.35	1e−15
CI-BCH 3	Cascade BCH	HD	6.7	9.35	1e−15
			12	9.90	
			20	10.3	
CI-BCH4	Cascade BCH	HD	6.7	9.55	1e−15
			12	10	
			20	10.5	
TPC	Iterative	HD/SD	7	9.3	1e−15
			15	11.1	
			20	11.4	

In Table 5.3, the schemes proposed during the preceding several years for extremely high-rate (more than 100 Gbps) optical communication systems of [52] are described. The schemes have a redundancy of about 7%. Most of these schemes make use of either LDPC codes or iterative/concatenated codes (including TPC; see Section 5.4), and all use hard decoding (denoted in the table by the abbreviation "HD").

Finally, Table 5.4 presents some of the characteristics of schemes for lower code rates (with a redundancy of about 20%). In this case, many of the same classes of codes are used, and most of them use soft decoding.

As seen in Table 5.4, all the coding schemes use either BCH codes, RS codes, LDPC codes, or combinations of these codes. Further development of coding systems for high-rate optical communication requires that the effect of the channel on the transmitted signal be taken into account. This can be achieved both by the standard approaches mentioned above, such as the turbo-equalizing method of multiplexing along with the orthogonal frequency division of channels (called OFDM, "orthogonal-frequency-division-multiplexing") – trying to "clean" the channel and cause it to behave like the ideal AWGN channel, as well as by using codes constructed to take the nature of the channel into account. There are many important issues that need to be properly stated and many open problems waiting for good solutions.

Table 5.4 Coding schemes for transmission of data at rates above 100 Gbps (with a redundancy of about 20%).

Title	Type	HD/SD	Redundancy (%)	NCG (dB)	BER
MTPC	Iterative RS	HD	20	9.3	1e−15
GLDPC	LDPC	HD	20	9.6	1e−15
TPC-BCH	Iterative BCH	HD	20	>10	1e−15
LDPC + SPC	Cascade	SD	20.5	11.3	1e−15
QC-LDPC	LDPC	SD	20	11.3	1e−15
CC-LDPC	Convolutional LDPC	SD	20	11.5	1e−15
LDPC + RS	Cascade	SD	20	9	1e−13
SC-LDPC	Cascade LDPC	SD	25.5	12	1e−15
NB-LDPC + RS	Cascade	SD	20.5	10.8	1e−15
NB-QC-LDPC	Non-double LDPC	SD	20	10.8	1e−12
3-concatenade NB-LDPC + RS	Cascade	SD	20.5	10.8	1e−15

References

1 Blahut, R. (1984). *Theory and Practice of Error Control Codes.* Reading, MA, USA: Addison-Wesley.

2 Blokh, E.L. and Zyablov, V.V. (1976). *Generic Cascade Codes (Algebraic Theory and Complicity in Realization).* Moscow: Svyaz (in Russian).

3 Blokh, E.L. and Zyablov, V.V. (1982). *Linear Cascade Codes.* Moscow: Nauka (in Russian).

4 Gallager, R.G. (1968). *Information Theory and Reliable Communication.* Chichester: Wiley.

5 Gallager, R. (1963). *Low-Densisity Parity-Check Codes.* Cambridge, MA, USA: MIT Press.

6 Zyablov, V.V. and Pinsker, M.S. (1975). Estimation of difficulties of errors correction by low-dense Gallagher's codes. (Russian) Zbl 0358.94017. *Probl. Peredachi Inf.* 11 (1): 23–26.

7 Kozlov, A.V., Krouk, E.A., and Ovchinnikov, A.A. (2013). Approach to block-exchanging codes design with low density of checking on evenness. *Izv. Vuzov, Instrum. Ind.* (8): 9–14. (in Russian).

8 Kolesnik, V.D. (2009). *Coding in Transmission, Transfer and Secure Information (Algebraic Theory of Block-Codes).* Moscow: Vishaya Shkola (in Russian).

9 Kolesnik, V.D. and Mironchikov, E.T. (1968). *Decoding of Cycle Codes*. Moscow: Svyaz (in Russian).

10 Kolesnik, V.D. and Poltirev, G.S. (1982). *Course of Information Theory*. Moscow: Nauka (in Russian).

11 Kudryashov, B.D. (2009). *Information Theory*. Piter: SPb (in Russian).

12 Mak-Vil'ams, A.D. and Sloen, D.A. (1979). *Codes Theory, Corrected Errors*. Moscow, Svyaz': (in Russian).

13 Piterson, U. and Ueldon, E. (1976). *Codes, Corrected Errors*. Moscow: Mir (in Russian).

14 Forni, D. (1970). *Cascade Codes*. Moscow: Mir in Russian.

15 Shannon, K. (1963). *Works on Information Theory and Cybernetics*. Inostrannya Literatura: Moscow (in Russian).

16 ITU-T Recommendation G.975/G709 (2000). *Forward error correction for submarine systems*.

17 ITU-T Recommendation G.975.1 (2004). *Forward error correction for high bit-rate DWDM submarine systems*.

18 Arabaci, M., Djordjevic, I.B., Saunders, R., and Marcoccia, R.M. (2010). Nonbinary quasi-cyclic LDPC based coded modulation for beyond 100-Gb/s. *IEEE Trans., Photonics Technol. Lett.* 22 (6): 434–436.

19 Bahl, L., Cocke, J., Jelinek, F., and Raviv, J. (1974). Optimal decoding of linear codes for minimizing symbol error rate. *IEEE Trans. Inform. Theory* IT-20 (2): 284–287.

20 Barg, A. and Zemor, G. (2005). Concatenated codes: serial and parallel. *IEEE Trans. Inform. Theory* 51 (5): 1625–1634.

21 Batshon, H.G., Djordjevic, I.B., Xu, L., and Wang, T. (2009). Multidimensional LDPC-coded modulation for beyond 400 Gb/s per wavelength transmission. *IEEE Photonics Technol. Lett.* 21 (16): 1139–1141.

22 Berlekamp, E., McEliece, R., and Van Tilborg, H. (1978). On the inherent intractability of certain coding problems. *IEEE Trans. Inform. Theory* 24 (3): 384–386.

23 Butler, B.K. and Siegel, P.H. (2014). Error floor approximation for LDPC codes in the AWGN channel. *IEEE Trans. Inform. Theory* 60 (12): 7416–7441.

24 Chen, J., Tanner, R.M., Jones, C., and Li, Y. (2005). Improved min-sum decoding algorithms for irregular LDPC codes. In: *Proceedings of International Symposium Information Theory (ISIT2005)*, 449–453. Adelaide, Australia, September.

25 Cole, C.A., Wilson, S.G., Hall, E.K., and Giallorenzi, T.R. (2006). Analysis and design of moderate length regular LDPC codes with low error floors. In: *Proceedings of 40th Annual Conference on Information Sciences and System (CISS 2006)*, (22–24 March). NJ, United States, 22-24 March: Princeton, pp. 823 and 828.

26 Diao, Q., Huang, Q., Lin, S., and Abdel-Ghaffar, K. (2012). A matrix-theoretic approach for analyzing quasi-cyclic low-density parity-check codes. *IEEE Trans. Inform. Theory* 58 (6), pp. 4030, 4048.

27 Djordjevic, I., Arabaci, M., and Minkov, L.L. (2009). Next generation FEC for high-capacity communication in optical transport networks. *J. Lightwave Technol.* 27, 15 (16): 3518–3530.

28 Djordjevic, I., Sankaranarayanan, S., Chilappagari, S.K., and Vasic, B. (2006). Low-density parity-check codes for 40-Gb/s optical transmission systems. *IEEE J. Sel. Top. Quantum Electron.* 12 (4): 555–562.

29 Djordjevic, I. and Vasic, B. (2005). Nonbinary LDPC codes for optical communication systems. *IEEE Photonics Technol. Lett.* 17 (10): 2224–2226.

30 Djordjevic, I., Xu, J., Abdel-Ghaffar, K., and Lin, S. (2003). A class of low-density parity-check codes constructed based on Reed–Solomon codes with two information symbols. *IEEE Photonics Technol. Lett.* 7 (7): 317–319.

31 Djordjevic, I., Xu, L., Wang, T., and Cvijetic, M. (2008). GLDPC codes with Reed–Muller component codes suitable for optical communications. *IEEE Commun. Lett.* 12 (9): 684–686.

32 Forestieri, E. (2005). *Optical Communication Theory and Techniques.* Germany: Springer.

33 Forney, G.D., Richardson, T.J., Urbanke, R.L., and Chung, S.-Y. (2001). On the design of low density parity-check codes within 0.0045 dB of the Shannon limit. *IEEE Commun. Lett.* 5 (2): 214–217.

34 Fossorier, M., Mihaljevic, M., and Imai, H. (1999). Reduced complexity iterative decoding of low density parity-check codes based on belief propagation. *IEEE Trans. Commun.* 47 (5): 379–386.

35 Guruswami, V. and Sudan, M. (1999). Improved decoding of Reed–Solomon and algebraic geometric codes. *IEEE Trans. Inform. Theory* IT-45 (6): 1755–1764.

36 Hirotomo, M. and Morii, M. (2010). Detailed evaluation of error floors of LDPC codes using the probabilistic algorithm. In: *Proceedings of International Symposium on Information Theory and Its Applications (ISITA2010)*, 513–518. Taichung, Taiwan, (17–20 October).

37 Gho, G.-H. and Kahn, J.M. (2012). Rate-adaptive modulation and low-density parity-check coding for optical fiber transmission systems. *IEEE/OSA J. Opt. Commun. Networking* 4 (10): 760–768.

38 Kabatiansky, G., Krouk, E., and Semenov, S. (2005). *Error Correcting Coding and Security for Data Networks: Analysis of the Super Channel Concept.* Wiley.

39 Kou, Y., Lin, S., and Fossorier, M.P.C. (2001). Low-density parity-check codes based on finite geometries: a rediscovery and new results. *IEEE Trans. Inform. Theory* 47 (7): 512–518.

40 Krouk, E. and Semenov, S. (eds.) (2011). *Modulation and Coding Techniques in Wireless Communications.* New Jersey: Wiley.

41 Li, J., Liu, K., Lin, S., and Abdel-Ghaffar, K. (2015). A matrix-theoretic approach to the construction of non-binary quasi-cyclic LDPC codes. *IEEE Trans. Commun.* 63 (4): 1057–1068.

42 Lin, S. and Ryan, W. (2009). *Channel Codes: Classical and Modern*. Cambridge: Cambridge University Press.

43 MacKay, D. (1999). Good error correcting codes based on very sparse matrices. *IEEE Trans. Inform. Theory* 45 (3): 363–368.

44 MacKay, D. and Neal, R. (2001). Near Shannon limit performance of low-density parity-check Codes. *IEEE Trans. Inform. Theory* 47 (2): 245–251.

45 Pyndiah, R. (1998). Near-optimum decoding of product codes: block turbo codes. *IEEE Trans. Commun.* 46 (8): 1003–1010.

46 Richardson, T.J. and Urbanke, R.L. (2001). The capacity of low-density parity-check codes under message-passing decoding. *IEEE Trans. Inform. Theory* 47 (2): 214–219.

47 Richardson, T.J. and Urbanke, R.L. (2001). Efficient encoding of low-density parity-check codes. *IEEE Trans. Inform. Theory* 47 (2): 278–285.

48 Richardson, T.J., Urbanke, R.L., and Chung, S.-Y. (2001). Analysis of sum-product decoding of low-density parity-check codes using a Gaussian approximation. *IEEE Trans. Inform. Theory* 47 (2): 239–245.

49 Shannon, C.E. (1948). A mathematical theory of communication. *Bell Syst. Tech. J.*: 29–35.

50 Schlegel, C. and Zhang, S. (2010). On the dynamics of the error floor behavior in (regular) LDPC codes. *IEEE Trans. Inform. Theory* 56 (7): 3248–3264.

51 Tanner, M. (1981). A recursive approach to low complexity codes. *IEEE Trans. Inform. Theory* 27: 533–547.

52 Vasic, B., Djordjevic, I., and Ryan, W. (2010). *Coding for Optical Channels*. Springer.

53 Xiao-Yu, H., Eleftheriou, E., and Arnold, D.M. (2005). Regular and irregular progressive edge-growth Tanner graphs. *IEEE Trans. Inform. Theory* 51 (1): 386–398.

54 Zhao, J., Zarkeshvari, F., and Banihashemi, A. (2005). On implementation of min-sum algorithm and its modifications for decoding low-density parity-check (LDPC) codes. *IEEE Trans. Commun.* 53 (4): 549–554.

6

Fading in Optical Communication Channels

As was mentioned in [1, 2], in most optical wired and wireless systems operated in the atmosphere [3, 4], there occur multiple diffraction, multiple reflection, and multiple scattering effects, due to artificial features inside fibers and dispersion of the inner material, or due to natural obstructions (aerosols, hydrometeors, such as rain, snow and clouds, etc.) filled in the atmosphere. All these effects, which cause not only addition losses (with respect to those obtained in line-of-sight [LOS] above-the-terrain conditions) but also addition fading of the signal strength observed at the detector, can be separated into two independent effects, the *slow* and the *fast* fading [5–10]. In real situations occurring in the atmospheric communication links, the various optical signals arriving at the moving or stationary detector via their respective paths have individual phases, which change continuously and randomly. Therefore, both the resultant signal envelope and its phase will also be random variables. Moreover, the Doppler effect can also be explained by random frequency modulation [11–14] due to different Doppler shifts of each of the received signals as multi-ray components. Therefore, to describe slow and fast signal fading, we need to introduce the corresponding mathematical description of the statistical effects that accompany real signal fading in multipath communication channels.

First, we will determine the main parameters of the multipath optical communication channel and the relations between channel parameters and those of the probing signal passing through it. Then, some specific statistical descriptions of the probability density function (PDF) and the cumulative distribution function (CDF) will be introduced based on well-known stochastic distributions and laws.

6.1 Parameters of Fading in Optical Communication Channel

In order to compare different optical multipath communication channels, wired and wireless, we should introduce a general receipt for designers of

Fiber Optic and Atmospheric Optical Communication, First Edition.
Nathan Blaunstein, Shlomo Engelberg, Evgenii Krouk, and Mikhail Sergeev.
© 2020 John Wiley & Sons, Inc. Published 2020 by John Wiley & Sons, Inc.

optical wired and wireless networks, of how to increase efficiency of such types of the multipath channel. First of all, we will describe the small-scale (in space domain) or fast (in time domain) variations of propagating optical signals, which, as was shown in [15–17], directly relate to the response of the channel under consideration. Later comes either a narrowband or a wideband (pulse) (see definitions in Chapter 3) that contains all information necessary to analyze and simulate any type of optical transmission through the channel. In modern optical communication links usually a wideband (pulse) signal is utilized and in our further discussions we mostly will talk about wideband wired or wireless optical channels and about their pulse response to optical signal with data propagation inside each specific channel.

Now, because of a time-varying impulse response due to optical source or optical detector (or obstructions surrounding) motion in the space domain, one can finally present a total received signal as a sum of amplitudes and time delays of the multi-ray components arriving at the detector at any instant of time, according to the similarity, described by a theorem of ergodicity of the processes, occurring both in time and space domains [13, 14]. If through measurements one can obtain information about the signal power delay profile, he finally can determine the main parameters of the optical multipath communication channel, both wired and wireless.

6.1.1 Time Dispersion Parameters

The first important parameters for wideband optical channels, which can be determined from a signal power delay profile, are *mean excess delay, rms delay spread*, and *excess delay spread* for the concrete threshold level X (in dB) of the channel.

The **mean excess delay** is the first moment of the power delay profile of the pulse signal and is defined using multi-ray signal presentation introduced in [5–7, 10]:

$$\langle \tau \rangle = \frac{\sum_{i=0}^{N-1} A_i^2 \tau_i}{\sum_{i=0}^{N-1} A_i^2} = \frac{\sum_{i=0}^{N-1} P(\tau_i)\tau_i}{\sum_{i=0}^{N-1} P(\tau_i)} \tag{6.1}$$

The *rms delay spread* is the square root of the second central moment of the power delay profile and is defined as

$$\sigma_\tau = \sqrt{\langle \tau^2 \rangle - \langle \tau \rangle^2} \tag{6.2}$$

where

$$\langle \tau^2 \rangle = \frac{\sum_{i=0}^{N-1} A_i^2 \tau_i^2}{\sum_{i=0}^{N-1} A_i^2} = \frac{\sum_{i=0}^{N-1} P(\tau_i)\tau_i^2}{\sum_{i=0}^{N-1} P(\tau_i)} \tag{6.3}$$

These delays are measured relative to the first detectable signal arriving at the optical detector at $\tau_0 = 0$. We must note that these parameters are defined from a single power delay profile, which was obtained after temporal or local (small-scale) spatial averaging of measured impulse response of the optical wireless channel. Usually, many measurements are made at many local (small-scale) areas in order to determine a statistical range of multi-ray parameters for optical communication networks over large-scale to medium-scale areas [5–7, 10, 14].

6.1.2   Coherence Bandwidth

As is known, the power delay profile in the time domain and the power spectral response in the frequency domain are related through the Fourier transform (see Chapter 3). Therefore, for a full description of the multi-ray optical channel the delay spread parameters in the time domain and the *coherence bandwidth* in the frequency domain are simultaneously used. The coherence bandwidth is the statistical measure of the frequency range over which the channel is considered "flat," i.e. a channel that passes all spectral multi-ray signals with equal gain and linear phase. In other words:

> Coherence bandwidth is a frequency range over which two frequency signals are strongly amplitude correlated.

This parameter actually describes the time disperse nature of the channel in a small-scale (local) area. Depending on the degree of amplitude correlation of two frequency separated signals, there are different definitions of this parameter.

The first definition is that the ***coherence bandwidth***, B_c, is a bandwidth over which the frequency correlation function is above 0.9 or 90%; then it equals

$$B_c \approx 0.02\sigma_\tau^{-1} \tag{6.4}$$

The second definition is that the ***coherence bandwidth***, B_c, is a bandwidth over which the frequency correlation function is above 0.5 or 50%; then it equals

$$B_c \approx 0.2\sigma_\tau^{-1} \tag{6.5}$$

It is important that there is no exact relationship between coherence bandwidth and rms delay spread; (6.5) and (6.6) are approximate. In general, one must use a special spectral analysis and more accurate multipath channel models [5–12].

6.1.3   Doppler Spread and Coherence Time

Above, we considered two parameters, *delay spread* and *coherence bandwidth*, which describe the time disperse nature of the multipath communication channel in a small-scale (local) area. To obtain information about the frequency dispersive nature of the channel caused by movements, either source/detector or surrounding obstructions, new parameters such as *Doppler spread* and *coherence time* are usually introduced to describe time variation phenomena of the channel in a small-scale region.

Doppler spread B_D is a measure that is defined as a range of frequencies over which the received Doppler spectrum is essentially nonzero. It shows the spectral spreading caused by the time rate of change of the mobile source or detector due to their relative motions (or scatterers surrounding them). According to [5–7, 10], Doppler spread B_D depends on Doppler shift f_D

$$f_D = \frac{v}{\lambda} \cos \alpha \tag{6.6}$$

And, via expression (6.6), on velocity of the moving detector/source v, the angle α between its direction of motion and direction of arrival of the reflected and/or scattered rays, and on the wavelength λ. If we deal with the complex baseband signal presentation (the definitions in Chapter 3), then we can introduce some criterion: if the baseband signal bandwidth is greater than the Doppler spread B_D, the effects of Doppler shift are negligible at the detector.

Coherence time T_c is the time domain dual of Doppler spread, and it is used to characterize the time-varying nature of the frequency non-dispersion of the channel in time coordinates. There is a simple relationship between these two channel characteristics, that is,

$$T_c \approx \frac{1}{f_m} = \frac{\lambda}{v} \tag{6.7}$$

We can also define the coherence time according to [6, 10, 14] as:

Coherence time is the time duration over which two multipath components of receiving signal have a strong potential for amplitude correlation.

If so, one can, as above for coherence bandwidth, define the time of coherence as the time over which the correlation function of two different signals in the time domain is above 0.5 (or 50%). Then, according to [5–7], we get

$$T_c \approx \frac{9}{16\pi f_m} = \frac{9\lambda}{16\pi v} = 0.18\frac{\lambda}{v} \tag{6.8}$$

As was shown in [10, 14], this definition is approximate and can be improved for modern digital communication channels by combining (6.7) and (6.8) as the geometric mean of them; that is,

$$T_c \approx \frac{0.423}{f_m} = 0.423\frac{\lambda}{v} \tag{6.9}$$

In the case described by (6.9), the correlation function of two different signals in the time domain is above 0.9 (or 90%). The definition of coherence time implies that two signals, which arrive at the receiver with a time separation greater than T_c, are affected differently by the channel.

6.2 Types of Small-Scale Fading

It is clear from channel parameter definitions that the type of optical signal fading within the multipath channel depends on the nature of the transmitting signal with respect to the characteristics of the channel. In other words, depending on the relation between the signal parameters, such as *bandwidth* B_S and *symbol period* T_S, and the corresponding channel parameters, such as *coherence bandwidth* B_c and *rms delay spread* σ_τ (or *Doppler spread* B_D and *coherence time* T_c), different transmitted signals will undergo different types of fading. As was shown in [8], there are *four possible scenarios* due to the time and frequency dispersion mechanisms in a mobile radio channel, which are manifested depending on the balance of parameters of the signal and of the channel mentioned above. The multipath time delay spread leads to *time dispersion* and *frequency-selective fading*, whereas Doppler frequency spread leads to *frequency dispersion* and *time-selective fading*. Separation between these four types of small-scale fading for impulse response of multipath radio channel is explained in Table 6.1 following and combining the material described in [6, 10]:

(1) *Fading due to multipath time delay spread*: Time dispersion due to multipath phenomena caused small-scale fading: either *flat* or *frequency-selective*:

a. The small-scale fading is characterized as *flat*, if the multipath optical channel has a constant-gain and linear-phase impulse response over a bandwidth that is *greater* than the bandwidth of the transmitted signal. Moreover, the signal bandwidth in the time domain (called the pulse or symbol period) exceeds the signal delay spread. These conditions are presented at the top rows of the last column of Table 6.1 both in the time and frequency domains. As follows from the illustrations presented, a flat fading channel can be defined as *narrowband* channel, since the bandwidth of the applied signal is *narrow* with respect to the channel flat fading bandwidth in the frequency domain, that is, $B_S < B_c$ (see Table 6.1). At the same time, the flat fading channel is an *amplitude-varying* channel, since a deep fading of the transmitted signal occurs within such a channel. Namely, such a channel corresponds to high obstructive conditions in the atmosphere, where, as will be shown below, strong multi-ray components of the received signal are present even in conditions of LOS.

b. The small-scale fading is characterized as *frequency selective*, if the multipath channel has a constant-gain and linear-phase impulse response over a bandwidth that is *smaller* than the bandwidth of the transmitted signal in the frequency domain, and its impulse response has a multiple delay spread greater than the bandwidth of the transmitted signal waveform (pulse or symbol period). These conditions are presented at the last column of Table 6.1 both in the time and frequency domains. As follows from Table 6.1, a *frequency-selective* fading channel can be defined as a *wideband* channel, since the bandwidth of the spectrum $S(f)$ of the transmitted signal is *greater* than the channel frequency-selective fading bandwidth in the frequency domain (the coherence bandwidth), that is, $B_S > B_c$ (see Table 6.1, the last column).

(2) *Fading due to Doppler spread*: Depending on how rapidly the transmitted baseband signal changes with respect to the rate of change of the channel, a channel may be classified as a *fast fading* or *slow fading* channel.

a. The channel in which the channel impulse response changes rapidly within the pulse (symbol) duration is called a *fast fading* channel. In other words, in such a channel its coherence time is smaller than the symbol period of the transmitted signal. At the same time, the Doppler spread bandwidth of the channel in the frequency domain is greater than the bandwidth of the transmitted signal (see the last column in Table 6.1). That is, $B_S < B_D$ and $T_c < T_S$. These effects cause frequency dispersion (also called *time-selective fading* [6, 10]) due to Doppler spreading, which leads to signal distortion. It is important to note that the fast-fading phenomenon occurs only due to source/detector motion.

In the case of flat-fading channel, one can model an impulse response by a delta function without any time delay spread. In this case, a *flat* and *fast-fading* channel corresponds to the channel in which the amplitude of the delta function varies *faster* than the rate of changes of the transmitted baseband signal $u(t)$ (see definitions in Chapter 3). In the case of a *frequency-selective* and *fast fading* channel, the amplitudes, phases, and time delays of any one of the multipath components vary *faster* than the rate of change of the transmitted bandpass signal $s(t)$ (see definitions in Chapter 3). In practical applications of optical communications the fast fading occurs only for vary low information rate.

b. The channel in which the channel impulse response changes at a rate *slower* than the transmitted baseband signal $u(t)$ is called a *slow-fading* channel. In this case, the channel may be assumed to be static over one or several bandwidth intervals. In the time domain, this implies that the reciprocal bandwidth of the signal is much smaller than the coherence time of the channel and in the frequency domain the Doppler spread of the channel is less than the bandwidth of the baseband signal, that is, $B_S \gg B_D$ and $T_S \ll T_c$. Both these conditions are presented at the bottom rows of the last column in Table 6.1. It is important to note that the velocity of the moving vehicle or moving obstructions within the channel as well as the baseband signal determines whether a signal undergoes

Table 6.1 Types of small-scale or fast fading (according to [10]).

General type of fading	Type of fading	Type of channel
		Narrowband
		(1.1.1): $B_S \ll B_c$
	(1.1): Flat fading	(1.1.2): $T_S \gg \sigma_\tau$
(1) Small-scale fading (Based on multipath time delay spread)		*Wideband*
	(1.2): Frequency selective Fading	(1.2.1): $B_S > B_c$
		(1.2.2): $T_S < \sigma_\tau$
		Narrowband
		(2.1.1): $B_S < B_D$
	(2.1): Fast fading	(2.1.2): $T_S > T_c$
(2) Small-scale fading (Based on Doppler frequency spread)		*Wideband*
	(2.2): Slow fading	(2.2.1): $B_S \gg B_D$
		(2.2.2): $T_S \ll T_c$

fast fading or slow fading. Thus, we summarize in Table 6.1 all the relation between the various multipath characteristics of the channel and the type of fading experienced by the desired signal.

6.3 Mathematical Description of Fast Fading

As was mentioned above, *fast* fading is observed both in situations when subscribers' detector or source are stationary and is caused by multiple reflections and scatterings from the surrounding moving obstructions, as well as when subscribers' detectors/sources are in motion and the fast fading at the detector is caused by multipath optical ray propagation, which occurs simultaneously with the fact that the source or the detector is in motion. In both situations the nature of the fast fading is the same, but is created by different sources. The relative motion of both terminals results, as was mentioned above, in the so-called Doppler shifts on each of the multipath wave components.

Now, we will discuss the question of the existence of a suitable statistical model for satisfactory description of multipath fast-fading channels. Several multipath models have been proposed to describe the observed random signal envelope and phase in a mobile channel. The earliest 2D (flat) model was presented by Clarke [11], which was based on the interference of direct (incident) wave and waves reflected from the obstruction. This model was briefly described in Chapter 3, where it was mentioned that application of such a 2D model is limited by the requirement of the existence of direct visibility between terminals, the source and the detector, and by a restricted range of reflection angles. In other words, this model cannot be used in real situations in the atmosphere where the direct path is almost always blocked by obstructions, such as clouds and rain, or other hydrometeors (see Chapters 10 and 11).

Now we will discuss the question of what PDFs and CDFs can be used to describe fast fading occurrence in the optical wireless multipath communication link. We will start with the worst case when a strong fading is observed in the optical wireless communication link, which, as was shown in [10–14], can be described more precisely by the use of Rayleigh PDF and CDF.

6.3.1 Rayleigh PDF and CDF

In optical wireless communication channels, stationary or dynamic, Rayleigh distribution is commonly used to describe the signal's spatial small-scale or temporal fast fading. As is known [10–14], a Rayleigh distribution can be obtained mathematically as the limit envelope of the sum of two quadrature Gaussian noise signals. Practically, this distribution can be predicted within the wireless communication channel if each multiray component of the received signal is independent. Here, once more, if the phase of multipath components is uniformly distributed over the range of $[0, 2\pi]$, then we deal with a zero-mean Rayleigh distribution of the envelope of the random variable x, the PDF of which can be presented in the following form:

$$\mathrm{PDF}(x) = \frac{x}{\sigma^2} \exp\left\{-\frac{x^2}{2\sigma^2}\right\}, \quad \text{for } x \geq 0 \tag{6.10}$$

where x is a signal envelope and σ is its standard deviation. The PDF distribution (6.10) completely describes the random received signal envelope $r(t)$ defined in 2D Clarke's model (see Chapter 3).

Here, the maximum value of $\mathrm{PDF}(x) = \exp(-0.5)/\sigma = 0.6065/\sigma$ corresponds to the random variable $x = \sigma$. According to Refs. [6, 10–14], one can also operate with the so-called *mean value*, the *rms value*, and the *median value* of the envelope of the random variable x. The definition of these parameters follows from the Rayleigh CDF presentation:

$$\mathrm{CDF}(X) = \Pr(x \leq X) = \int_0^X \mathrm{PDF}(x)dx = 1 - \exp\left\{-\frac{X^2}{2\sigma^2}\right\} \tag{6.11}$$

All parameters in (6.11) are the same, as for (6.10). The mean value of the Rayleigh distributed signal strength (voltage), x_{mean} (in the literature it is also denoted as $E[x]$), can be obtained from the following conditions:

$$x_{\mathrm{mean}} \equiv E[x] = \int_0^\infty x \cdot \mathrm{PDF}(x)dx = \sigma \cdot \sqrt{\frac{\pi}{2}} \approx 1.253\sigma \tag{6.12}$$

At the same time, the variance or average power of the received signal envelope for the Rayleigh distribution can be determined as

$$\sigma_x^2 \equiv E[x^2] - E^2[x] \equiv \langle x^2 \rangle - \langle x \rangle^2$$

$$= \int_0^\infty x^2 \cdot \text{PDF}(x)dx - \frac{\pi\sigma^2}{2} = \sigma^2 \left(2 - \frac{\pi}{2}\right) \approx 0.429\sigma^2 \tag{6.13}$$

If so, the *rms value* of the signal envelope is defined as the square root of the mean square, that is,

$$\text{rms} = \sqrt{2} \cdot \sigma \approx 1.414\sigma \tag{6.14}$$

The *median value* of Rayleigh distributed signal strength envelope is defined from the following conditions:

$$\frac{1}{2} = \int_0^{x_{\text{median}}} \text{PDF}(x)dx \tag{6.15}$$

from which it immediately follows that

$$x_{\text{median}} = 1.177\sigma \tag{6.16}$$

All these parameters are presented on the x-axis in Figure 6.1. As follows from (6.12), (6.14), and (6.16), the difference between the mean and the median values is ~0.076σ and their PDF for a Rayleigh fading signal envelope differs by only 0.55 dB. The differences between the *rms* value and two other values are higher. We must note here that in the practice of wireless systems' servicing, the *median value* is often used, since fading data are usually measured in the wave field arriving at the receiver in real time.

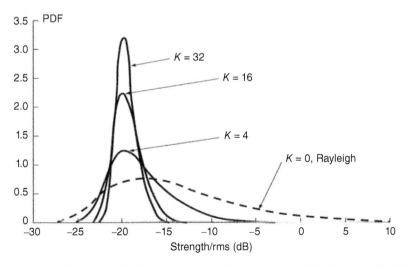

Figure 6.1 Ricean PDF distribution vs. (x/rms) for various K = 0, 4, 16, 32 (according to [10]).

6.3.2 Ricean PDF and CDF

Usually, in wireless communication links, radio and optical, not only multi-ray components arrive at the detector due to multiple reflection, diffraction, and scattering from various obstructions surrounding the two terminals, the source and the detector. An LOS component, which describes signal loss along the path of direct visibility (called the *dominant path* [6, 10–14]) between both terminals, is also often found at the detector. This dominant component of the received signal may significantly decrease the depth of interference picture. The PDF of such a received signal is usually said to be *Ricean*. To estimate the contribution of each component, dominant (or LOS) and multipath, for the resulting signal at the receiver, the Ricean parameter K is usually introduced as a ratio between these components, i.e.

$$K = \frac{\text{LOS-component power}}{\text{Multipath-component power}} \qquad (6.17)$$

It must be noted that the Ricean distribution consists of all the other distributions mentioned above. Thus when $K = 0$, that is, not any dominant component but only scattered and reflected components occur at the receiver, the communication channel is Rayleigh with the PDF described by (6.10). In the other limiting case of $K \to \infty$ the propagation channel is limited to Gaussian PDF that will be shown below, as a law describing slow-fading phenomena or ideal multipath channel with flat fast fading.

Thus, we can define the Ricean distribution as the general case of signal fading along any optical path between two terminals in optical wireless communication link. As the dominant component at the receiver becomes weaker, the resulting signal is closer to that obtained after multiple reflections and scattering, which has an envelope that is Rayleigh. In this case, the fades with a high probability are very deep, whereas if K limits to infinity the fades are very shallow. The above mentioned is illustrated in Figure 6.1, where a set of *PDF*s, which describe the envelope fading profiles for different values of K vs. the signal level, is presented, normalized to the corresponding local mean.

The Ricean *PDF* distribution of the signal strength or voltage envelope, denoted by arbitrary random variable x, can be defined as [6, 10–14]

$$\text{PDF}(x) = \frac{x}{\sigma^2} \exp\left\{-\frac{x^2 + A^2}{2\sigma^2}\right\} \cdot I_0\left(\frac{Ax}{\sigma^2}\right), \quad \text{for } A > 0, \ x \geq 0 \quad (6.18)$$

where A denotes the peak strength or voltage of the dominant component envelope and $I_0(\cdot)$ is the modified Bessel function of the first kind and zero order. According to definition (6.18), one can now rewrite the parameter K that was defined above as the ratio between the *dominant* and the *multipath* component power. It is given by

$$K = \frac{A^2}{2\sigma^2} \qquad (6.19)$$

or in terms of dB

$$K = 10 \log \frac{A^2}{2\sigma^2} \text{ (dB)} \tag{6.20}$$

Since the Ricean CDF is also of interest, we define this CDF as

$$\text{CDF}(X) = \text{Pr}(x \leq X) = \int_{-\infty}^{X} \text{PDF}(x)dx = \frac{1}{2}\left[1 + \text{erf}\left(\frac{X}{\sqrt{2} \cdot \sigma}\right)\right] \tag{6.21}$$

It can be shown that the Ricean CDF can be rewritten by the quadrature formula according to [16] instead of that presented in a very complicated form (6.21).

In fact, if we introduce the Ricean K-factor from (6.19) we first rewrite (6.18) as a function only of K:

$$\text{PDF}(x) = \frac{x}{\sigma^2} \exp\left\{-\frac{x^2}{2\sigma^2}\right\} \cdot \exp(-K) \cdot I_0\left(\frac{x}{\sigma}\sqrt{K}\right) \tag{6.22}$$

for $K = 0 \exp(-K) = 1$ and $I_0(*) = 1$ the worst case Rayleigh PDF follows when there is no dominant signal component, that is, the distribution described by (6.11). Conversely, in a situation of direct vision between two terminals with no multipath components, that is $K \to \infty$, the Ricean fading approaches a Gaussian one yielding a "Dirac-delta shaped" PDF (see Figure 6.2). Hence, if the K-factor is known, the signal fading envelope's distributions (PDF and CDF) are described perfectly.

It should be noted that the fast-fading distribution is usually described by the Ricean PDF, which is introduced by Eq. (6.18) or (6.22). The Ricean PDF limits to the Rayleigh PDF and to the Gaussian PDF, if the Ricean K-factor limits to zero or infinity, respectively.

Additionally, it should be pointed out that Ricean CDF has a much more difficult form of presentation with respect to those of Rayleigh CDF (6.19) and Gaussian CDF (6.22), because it is presented by the sum of infinite number of products of modified Bessel functions $I_m(\cdot)$ of the m-order:

$$\text{CDF}(X) = 1 - \exp\left\{-\left(K + \frac{x^2}{2\sigma_r^2}\right)\right\} \cdot \sum_{m=0}^{\infty}\left(\frac{\sigma_r\sqrt{2K}}{r}\right) \cdot I_m\left(\frac{x \cdot \sqrt{2K}}{\sigma_r}\right) \tag{6.23}$$

On practice, for computations of (6.23), in the sum $\sum(*)$ it usually used only that summand, which gives in sum additional effect of 0.1%.

In [18] instead of the complicated formula (6.23) a quadrature formula was obtained that allows to present the mean value and its standard deviation in very simple "engineering" form based on the fade parameter K. Thus, in [18] the mean value μ_r and the variation $\sigma_r^2(K)$ of the signal envelope r of varied signal

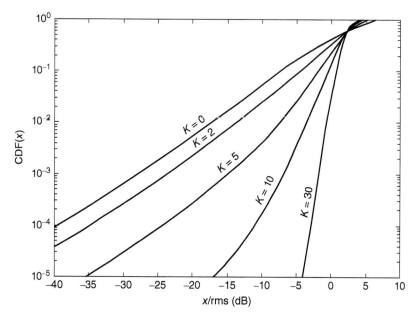

Figure 6.2 CDF in logarithmic scale vs. (x/rms) for various $K = 0, 2, 5, 10, 30$ (according to [10]).

amplitude x, caused by fast fading, respectively, were obtained in the following manner:

$$\mu_r \equiv \bar{x}(K) = \int_0^\infty x \cdot \text{PDF}(x)dx = [(1 + K)I_0(\sqrt{2Kx}) + KI_1(K/2)] \quad (6.24)$$

and

$$\sigma_r^2(K) = \int_0^\infty x^2 \cdot \text{PDF}(x)dx = 2 \cdot (1 + K) - \mu_r^2 \quad (6.25)$$

Here, as above, $I_1(\cdot)$ – is a modified Bessel function of the first order.

In practical point of view, when $K \geq 2$ the effect of multi-ray and the corresponding fast fading become weaker, we obtain, following [18], a simple formula for the mean value and the variance of signal envelope r; that is,

$$\mu_r \equiv \bar{x}(K) = \sqrt{2K}\left(1 - \frac{1}{4K} + \frac{1}{K^2}\right) \quad (6.26a)$$

and

$$\sigma_r^2(K) = 1 - \frac{1}{4K}\left(1 + \frac{1}{K} - \frac{1}{K^2}\right) \quad (6.26b)$$

As in a situation of strong fading occurring in the optical channel, when $K < 2$, we obtain, following [18], more complicated expressions for the mean and the variance of signal envelope r in the Ricean channel with fast fading, respectively:

$$\mu_r \equiv \overline{x}(K) = \sqrt{\frac{\pi}{2} + \sum_{n=1}^{\infty} \frac{2}{\pi} \cdot \frac{(-1)^n}{(2n-1)!} K^n} \tag{6.27a}$$

$$\sigma_r^2(K) = 1 - \left(\frac{\pi}{2} - 1\right) \cdot \exp\left\{-\frac{K}{\sqrt{2}}\right\} \tag{6.27b}$$

It is clearly seen that for a known parameter of fast fading K (called in literature the Ricean parameter), the distribution of the signal envelope r can be strictly estimated and then, its PDF and CDF. In other words, it is easy to estimate the influence of fast fading on optical signal propagating in the real optical channel, wired or wireless.

6.3.2.1 Gamma-Gamma Distribution

The more common and well-known model to characterize the irradiance fluctuations caused in atmospheric wireless communication links that take into account also the assumed modulation processes of optical wave propagation is the Gamma-Gamma distribution model [15–17]. The Gamma-Gamma distribution can be derived from the modulation process, where the double stochastic negative exponential distribution is directly related to the atmospheric parameters, by describing the large- and the small-scale obstructions randomly distributed along the optical ray trajectory by *Gamma* distributions [15–17]:

$$p_x(x) = \frac{\alpha(\alpha x)^{\alpha-1}}{\Gamma(\alpha)} \exp(-\alpha x), \quad \text{for } x > 0, \ \alpha > 0 \tag{6.28a}$$

$$p_y(y) = \frac{\beta(\beta y)^{\beta-1}}{\Gamma(\beta)} \exp(-\beta y), \quad \text{for } y > 0, \ \beta > 0 \tag{6.28b}$$

where x and y are the mean irradiance, $x = y = \langle I \rangle$ are random quantities, and I is the irradiance. In (6.28a) and (6.28b) present Gamma functions of parameters α and β, respectively, and first, by fixing x and then writing $y = I/x$, we obtain the conditional *PDF*:

$$p_y(I \mid x) = \frac{\beta(\beta I/x)^{\beta-1}}{x\Gamma(\beta)} \exp(-\beta I/x), \quad \text{for } I > 0 \tag{6.29}$$

$$p(I) = \int_0^{\infty} p_y(I \mid x) p_x(x) dx$$

$$= \frac{2(\alpha\beta)^{(\alpha+\beta)/2}}{\Gamma(\alpha)\Gamma(\beta)} I^{[(\alpha+\beta)/2]-1} K_{\alpha-\beta}[2(\alpha\beta I)^{1/2}], \quad \text{for } I > 0 \tag{6.30}$$

where $K_{\alpha-\beta}[\cdot]$ represents the Bessel function of the second kind of $(\alpha - \beta)$ order.

This *Gamma-Gamma* distribution can be divided into two regimes, considering parameters α and β, where α represents effective number of the large-scale random variables and β is related to that of the small-scale random variables. When signal fading is weak (i.e. the atmospheric channel is not strongly renormalized), effective numbers of scale sizes are either smaller or much larger relative to the first Fresnel zone (see [10, 15–17]).

Strong variations of atmospheric channel parameters can be characterized by β decreasing beyond the focusing regime and approaching saturation, where $\beta \to 1$ means that the number of small-scale obstructions is reduced to one. On the other hand, the number of discrete obstructions, α, increases under strong fading conditions [15–17].

Nevertheless, the *Gamma-Gamma* distribution can be approached as a negative exponential distribution [15–17]. The large- and the small-scale parameters of mean fading may be written as

$$\alpha = \frac{1}{\sigma_x^2}, \quad \beta = \frac{1}{\sigma_y^2} \tag{6.31}$$

where σ_x^2 and σ_y^2 are the normalized variances of x and y, respectively. Eventually, the total scintillation index is (see definitions introduced in Chapter 11 for a turbulent atmospheric communication links)

$$\sigma_I^2 = \frac{1}{\alpha} + \frac{1}{\beta} + \frac{1}{\alpha\beta} \tag{6.32}$$

This distribution will be used for analyzing data stream parameters estimation in wireless atmospheric communication links.

6.4 Mathematical Description of Large-Scale Fading

As was mentioned above, the random effects of shadowing or diffraction from obstructions result in *slow* (in time domain) or *large-scale* (in space domain) fading. Because in decibels this phenomenon is described by normal or Gaussian distribution, in watts (or voltage) it can be described by the use of the lognormal distribution [8]. As was expected, a lognormal distribution in terms of decibels was obtained due to normally distributed random shadowing effects, that is, due to diffraction from randomly distributed obstructions placed in the atmosphere along the optical ray trajectory (see Chapter 10). Such a distribution follows from normal or Gaussian, distribution of obstruction usually obtained in experiments for non-line-of-sight (NLOS) urban microcell channels. Moreover, it is possible to have an LOS condition with absence of multipath, giving a Gaussian or the lognormal distribution.

6.4.1 Gaussian PDF and CDF

A very interesting situation within the communication link is that when *slow* signal fading occurs at the detector it has tendency to lognormal or Gaussian distribution of the amplitude of the PDF [5–7, 10–14]. According to [10, 14], we define Gaussian or normal distribution of the envelope of the random variable *x* (let us say, the optical signal strength or voltage) by introducing the following PDF of the envelope of the received signal random *x*:

$$\text{PDF}(x) = \frac{1}{\sigma\sqrt{2\pi}} \exp\left\{ -\frac{(x - \bar{x})^2}{2\sigma^2} \right\} \tag{6.33}$$

Here $\bar{x} \equiv \langle x \rangle$ is the mean value of the random signal level, σ is the value of the received signal strength or voltage deviations, and $\sigma^2 = \langle x^2 - \bar{x}^2 \rangle$ is the variance or time-average power ($\langle w \rangle$ is a sign of averaging of variable *w*) of the received signal envelope. This PDF can be obtained only as a result of the random interference of a large number of signals with randomly distributed amplitudes (strength or voltage) and phase. If the phase of the interfering signals is uniformly distributed over the range of $[0, 2\pi]$, then one can talk about a zero-mean Gaussian distribution of random variable *x*. In this case, we define the PDF of such a process by (6.33) with $\bar{x} = 0$ and $\sigma^2 = \langle x^2 \rangle$ in it. The definition of the mean value and the variance of random signal envelope is also used for description of the so-called CDF. This function describes the probability of the event that the envelope of received signal strength (voltage) does not exceed a specified value *X*, that is,

$$\text{CDF}(X) = \Pr(x \leq X) = \int_0^X \text{PDF}(x)dx$$

$$= \frac{1}{\sigma\sqrt{2\pi}} \int_{-\infty}^X \exp\left\{ -\frac{(x - \bar{x})^2}{2\sigma^2} \right\} = \frac{1}{2} + \frac{1}{2}\text{erf}\left\{ \frac{X - \bar{x}}{\sqrt{2}\sigma} \right\} \tag{6.34}$$

where the error function is defined by

$$\text{erf}(w) = \frac{2}{\sqrt{\pi}} \int_0^w \exp(-y^2)dy \tag{6.35}$$

Using distributions (6.33), (6.34), and knowledge about the mean value $\bar{x}$ (if it is not zero) and about the variance $\sigma^2 = \langle x^2 \rangle$ of received signals, as well as the data obtained during measurements of signal amplitude *x*, one can easily predict the Gaussian fading as a source of the multiplicative noise effects.

References

1 Marcuse, O. (1972). *Light Transmission Optics*. New York: Van Nostrand-Reinhold Publisher.

2 Midwinter, J.E. (1979). *Optical Fibers for Transmission*. New York: Wiley.

3 Djuknic, G.M., Freidenfelds, J., and Okunev, Y. (1997). Establishing wireless communication services via high-altitude aeronautical platforms: a concept whose time has come? *IEEE Communication Magazine*: 128–135.

4 Hase, Y., Miura, R., and Ohmori, S. (1998). A novel broadband all-wireless access network using stratospheric radio platform, *VTC'98 (48th Vehicular Technology Conference)*, Ottawa, Canada, May, 1998.

5 Lee, W.Y.C. (1985). *Mobile Communication Engineering*. New York: McGraw-Hill Publications.

6 Rappaport, T.S. (1996). *Wireless Communications*. New York: Prentice Hall PTR.

7 Stremler, F.G. (1982). *Introduction to Communication Systems*. Addison-Wesley Reading.

8 Biglieri, E., Proakis, J., and Shamai, S. (1998). Fading channels: information theoretic and communication aspects. *IEEE Trans. Inform. Theory* 44 (6): 2619–2692.

9 Nielsen, T.H. (2001). IPv6 for wireless networks. *J. Wireless Personal Commun.* 17: 237–247.

10 Blaunstein, N. and Christodoulou, C. (2007). *Radio Propagation and Adaptive Antennas for Wireless Communication Links: Terrestrial, Atmospheric and Ionospheric*. New Jersey: Wiley InterScience.

11 Clarke, R.H. (1968). A statistical theory of mobile-radio reception. *Bell Syst. Tech. J.* 47: 957–1000.

12 Aulin, T. (1979). A modified model for the fading signal at a mobile radio channel. *IEEE Trans. Veh. Technol.* 28 (3): 182–203.

13 Papaulis, A. (1991). *Probability, Random Variables, and Stochastic Processes*. New York: McGraw-Hill.

14 Proakis, J.G. (1995). *Digital Communications*. New York: McGraw-Hill.

15 Andrews, L.C., Phillips, R.L., and Hopen, C.Y. (2001). *Laser Beam Scintillation with Applications*. Bellingham, WA: SPIE Optical Engineering Press.

16 Andrews, L.C. (1998). *Special Functions of Mathematics for Engineers*, 2e. Bellingham/Oxford: SPIE Optical Engineering Press/Oxford University Press.

17 Andrews, L.C. and Phillips, R.L. (1998). *Laser Beam Propagation through Random Media*. Bellingham, WA: SPIE Optical Engineering Press.

18 Krouk, E. and Semionov, S. (eds.) (2011). *Modulation and Coding Techniques in Wireless Communications*. Chichester, England: Wiley.

7

Modulation of Signals in Optical Communication Links

As was shown in Chapter 3, there are two main types of optical signals propagating in wired or wireless communication links: time continuously varied or
analog, which corresponds to narrowband channels, and time discrete varied
or pulse-shaped, which corresponds to wideband channels [1–5]. Therefore,
there are different types of modulation that are usually used for such types of
signals. First, we will define the process of modulation and demodulation.

Modulation is the process where the message information is added to the
radio carrier. In other words, modulation is the process of encoding information from a message source in a manner suitable for transmission. This process
involves translating a baseband message signal, the source, to a bandpass signal
at frequencies that are very high with respect to the baseband frequency (see
definitions in Chapter 3). The bandpass signal is called the *modulated* signal
and the baseband message signal is called the *modulating* signal [3–10].

Modulation can be done by varying the amplitude, phase, or frequency of a
high-frequency carrier in accordance with the amplitude of the baseband message signal. These kinds of analog modulation have been employed in the first
generation of wireless systems and have continued until today for LIDAR and
optical imaging applications. Further, digital modulation has been proposed
for use in present day radio and optical communication systems. Because this
kind of modulation has numerous benefits compared with conventional analog modulation, the primary emphasis of this topic is on digital modulation
techniques and schemes (see the next Section 7.2). However, since analog modulation techniques are in widespread use till date and will continue in the future,
they are considered first.

Demodulation is the process of extracting the baseband message from the
carrier so that it may be processed and interpreted by the intended radio or
optical receiver [1–3]. Since the main goal of a modulation technique is to transport the message signal through an optical communication channel, wired or

Fiber Optic and Atmospheric Optical Communication, First Edition.
Nathan Blaunstein, Shlomo Engelberg, Evgenii Krouk, and Mikhail Sergeev.
© 2020 John Wiley & Sons, Inc. Published 2020 by John Wiley & Sons, Inc.

wireless, with the best possible quality while occupying the least amount of frequency band spectrum, many modern practical modulation techniques have been proposed to increase the quality and efficiency of various optical communication links, including fiber optical links.

Below, we will briefly describe the main principle of both kinds of modulation, analog and digital, and will give some examples of the most useful types of modulation adapted for both kinds of channels, narrowband and wideband.

7.1 Analog Modulation

Each analog signal consists of three main time-varying characteristics: the amplitude $a(t)$, the phase $\phi(t)$, and the angular frequency $\omega(t) = 2\pi f(t)$, since there is a simple relation between the phase and the frequency $\phi(t) = \omega(t) \cdot t + \phi_0$, where ϕ_0 is the initial phase of the signal. In other words, any signal can be presented via these three parameters as

$$x(t) = a(t)e^{j\phi(t)} = a(t)e^{j[\omega(t)\cdot t + \phi_0]} \tag{7.1}$$

Consequently, there are three types of modulation, depending on what characteristic is time varied in the modulating signal (called *message*; see the definitions above) – amplitude modulation (AM), phase modulation (PM), and frequency modulation (FM).

7.1.1 Analog Amplitude Modulation

In AM technique, the amplitude of a high-frequency carrier signal is varied in accordance to the instantaneous amplitude of the modulating message signal. The AM signal can be represented through the carrier signal and the modulating message signal as

$$s_{AM}(t) = A_c[1 + m(t)]\cos(2\pi f_c t) \tag{7.2}$$

where $x_c(t) = A_c\cos(2\pi f_c t)$ is a carrier signal with amplitude A_c and high frequency f_c, and $m(t) = (A_m/A_c)\cos(2\pi f_m t)$ is a sinusoidal modulating signal with amplitude A_m and low frequency f_m. Usually, the modulation index $k_m = (A_m/A_c)$ is introduced, which is often expressed in percentage and is called *percentage modulation*. Figure 7.1 shows a sinusoidal modulating signal $m(t)$ and the corresponding AM signal $s_{AM}(t)$ for the case $k_m = (A_m/A_c) = 0.5$, that is, the signal is said to be 50% modulated.

If $k_m(\%) > 100\%$, the message signal will be distorted at the envelope detector. Equation (7.2) can be rewritten as

$$s_{AM}(t) = \text{Re}[g(t)\exp(j2\pi f_c t)] \tag{7.3}$$

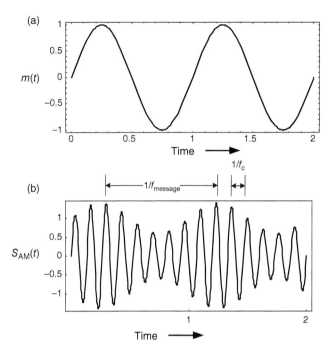

Figure 7.1 The amplitude modulating (a) and modulated (b) signals for the modulation index $k_m = 0.5$.

where $g(t)$ is the complex envelope of the AM signal given according to (7.2) by

$$g(t) = A_c[1 + m(t)] \tag{7.4}$$

The corresponding spectrum of an AM signal can be shown to be [1–4]

$$S_{AM}(f) = \frac{1}{2}A_c[\delta(f - f_c) + S_M(f - f_c) + \delta(f + f_c) + S_M(f + f_c)] \tag{7.5}$$

where $\delta(\cdot)$ is the Delta-function, and $S_M(f)$ is the message signal spectrum.

The bandwidth of an AM signal is equal to $B_{AM} = 2f_m$ where f_m is the maximum frequency contained in the modulating message signal. The total power in an AM signal can be obtained as [1–4]

$$P_{AM}(t) = \frac{1}{2}A_c[1 + 2\langle m(t)\rangle + \langle m^2(t)\rangle] \tag{7.6}$$

where $\langle m(t)\rangle$ represents the average value of the message signal. Using the expression of the message signal through the modulation index presented above, one can simplify expression (7.6) as

$$P_{AM}(t) = \frac{1}{2}A_c[1 + P_m] = P_c\left[1 + \frac{k_m^2}{2}\right] \tag{7.7}$$

where $P_c = \frac{1}{2}A_c^2$ is the power of the carrier signal and $P_m = \langle m^2(t) \rangle$ is the power of the modulating message signal. It can be shown that

$$\frac{1}{2}[P_{AM} - P_c] = \frac{1}{2}\left[P_c + P_c\frac{k_m^2}{2} - P_c\right] = \frac{1}{2}\left[P_c\frac{k_m^2}{2}\right] = \frac{1}{2}\left[\frac{A_c^2}{2}\frac{A_m^2}{2A_c^2}\right] = \frac{1}{8}A_m^2 = \frac{1}{4}P_m \tag{7.8}$$

from which it follows that $\lfloor P_{AM} - P_c \rfloor = P_m/2$.

7.1.2 Analog Angle Modulation – Frequency and Phase

FM is a part of an *angle modulation* technique where the instantaneous frequency of the carrier, $f_c(t)$, varies linearly with the baseband modulating waveform, $m(t)$, i.e.

$$f_c(t) = f_c + k_f m(t) \tag{7.9}$$

where k_f is the frequency sensitivity (the frequency deviation constant) of the modulator measured in Hz/V. To understand what it means, let us first explain the *angle modulation* technique.

Angle modulation varies a sinusoidal carrier signal in such a way that the phase θ of the carrier is varied according to the amplitude of the modulating baseband signal. In this technique of modulation, the amplitude of the carrier wave is kept constant (called the *constant envelope* modulation). There are several techniques to vary the phase $\theta(t)$ of a carrier signal in accordance with the baseband signal. The well-used techniques of angle modulation are the FM and the PM. In *FM signal* the instantaneous carrier phase is

$$\theta(t) = 2\pi \int_0^t f_c(t')dt' = 2\pi \left[f_c t + k_f \int_0^t m(t')dt'\right] \tag{7.10}$$

If so, the bandpass FM signal can be presented in the following form:

$$s_{FM}(t) = \text{Re}[g(t)\exp(j2\pi f_c t)] = A_c \cos\left[2\pi f_c t + 2\pi k_f \int_0^t m(t')dt'\right] \tag{7.11}$$

Here, the envelope $g(t)$ is the complex lowpass FM signal:

$$g(t) = A_c \exp\left[2\pi k_f \int_0^t m(t')dt'\right] \tag{7.12}$$

where as before, $\text{Re}[w]$ is the real part of w. The process of FM is illustrated in Figure 7.2.

Note that FM is a constant envelope modulation technique, making it suitable for nonlinear amplification. If, for example, the modulating baseband signal is sinusoidal of amplitude and frequency, then the FM signal can be expressed as

$$s_{FM}(t) = A_c \cos\left[2\pi f_c t + \frac{k_f A_m}{f_m}\sin(2\pi f_m t)\right] \tag{7.13}$$

Figure 7.2 The modulating signal (a), time-varied modulation frequency (b), and FM signal (c).

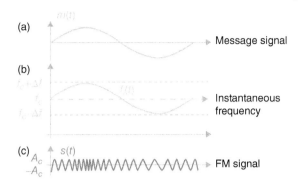

(a) Message signal

(b) Instantaneous frequency

(c) FM signal

7.1.2.1 Phase Modulation

In PM signal the angle $\theta(t)$ of the carrier signal is varied linearly with the baseband message signal $m(t)$, and can be presented in the same manner as FM signal, that is,

$$s_{\text{PM}}(t) = A_c \cos[2\pi f_c t + k_\theta m(t)] \tag{7.14}$$

In (7.14) k_θ is the phase sensitivity of the modulator (the phase deviation constant) measured in *radians per volt*. From (7.11) and (7.14) it follows that an FM signal can be regarded as a PM signal in which the lowpass modulating wave is integrated before modulation. So, an FM signal can be generated by first integrating $m(t)$ and then using the result as an input to a phase modulator. Conversely, a PM signal can be generated by first differentiating $m(t)$ and then using the result as the input to a frequency modulator.

The FM index defines the relationship between the message amplitude and the bandwidth of the transmitted signal, which is presented in the following form:

$$\beta_f = \frac{k_f A_m}{B_f} = \frac{\Delta f}{B_f} \tag{7.15}$$

where, as above, A_m is the peak value of the modulating message signal, Δf is the peak frequency deviation of the transmitter, and B_f is the maximum bandwidth of the modulating lowpass signal (usually B_f is equal to the highest frequency component f_m presented in the modulating signal and simply $\beta_f = \Delta f / f_m$).

The PM index is defined as

$$\beta_\theta = k_\theta A_m = \Delta\theta \tag{7.16}$$

where $\Delta\theta$ is the peak phase deviation of the transmitter.

7.1.3 Spectra and Bandwidth of FM or PM Signals

Since PM and FM signals have the same form of presentation of modulated signal, we will pay attention to one of them, let us say an FM signal. An FM signal is a nonlinear function of the modulating waveform $m(t)$ and, there-fore, the spectral characteristics of $s(t)$ cannot be obtained directly from the spectral characteristics of $m(t)$. However, the bandwidth of $s(t)$ depends on $\beta_f = \Delta f / f_m$. If $\beta_f \gg 1$, then the narrowband FM signal is generated, where the spectral widths of $s(t)$ and $m(t)$ are about the same, i.e. $2f_m$. If $\beta \ll 1$, then the wideband FM signal is generated where the spectral width of $s(t)$ is slightly greater than $2\Delta f$. For arbitrary FM index, the approximate bandwidth of the FM signal (in which this signal has 98% of the total power of the transmitted RF signal), which continuously limits to these upper and lower bounds, is [3–5]

$$B_T = \begin{cases} 2\Delta f \left(1 + \frac{1}{\beta_f}\right) \approx 2(\beta_f + 1)f_m, & \beta \ll 1 \\ 2\Delta f \left(1 + \frac{1}{\beta_f}\right) \approx 2\Delta f, & \beta \gg 1 \end{cases} \tag{7.17}$$

This approximation of FM bandwidth is known as *Carson's rule* [3–5]. It states that for the upper bound, the spectrum of FM signal is limited to the carrier frequency f_c of the carrier signal, and one pair of sideband frequencies at $f_c \pm f_m$. For the lower bound, the spectrum of FM signal is simply slightly greater than $2\Delta f$.

There are two variants of FM signal generation, the direct and the indirect, as well as many methods of its demodulation by using different kinds of detectors. This specific subject is out of the scope of the current book; therefore, we refer the reader to special literature [1–5], where these questions are fully described.

7.1.4 Relations Between SNR and Bandwidth in AM and FM Signals

In angle modulation systems, the signal-to-noise ratio (SNR) *before* detection is a function of the receiver intermediate frequency (IF) filter bandwidth (see formulas above, where optical signal AM and FM modulation is described) of the received carrier power, and of the received interference [3–5]; that is,

$$(\text{SNR})_{\text{in}} = \frac{A_c^2/2}{2N_0(\beta_f + 1)B_F} \tag{7.18}$$

where A_c is the carrier amplitude, N_0 is the white noise RF power spectral density (PSD), and B_F is the equivalent RF bandwidth of the bandpass filter at the front end of the receiver. Note that $(\text{SNR})_{\text{in}}$ uses the carrier signal bandwidth according to Carson's rule (7.17).

However, the SNR *after* detection is a function of the maximum frequency of the message, f_m, the modulation index, β_f or β_θ, and the given SNR at the input of the detector, $(\text{SNR})_{\text{in}}$. For example, the SNR at the output of an FM receiver depends on the modulation index and is given by [5]

$$(\text{SNR})_{\text{out}} = 6(\beta_f + 1)\beta_f^2 \left\langle \left(\frac{m(t)}{V_p}\right)^2 \right\rangle (\text{SNR})_{\text{in}} \tag{7.19}$$

where V_p is the peak-to-zero value of the modulating signal $m(t)$.

For comparison purposes, let us present here the $(SNR)_{in}$ for an AM signal, which, according to [3], is defined as the input power to a conventional AM receiver having RF bandwidth equal to $2B_F$; that is,

$$(SNR)_{in,AM} = \frac{P_c}{N} = \frac{A_c^2}{2N_0 B_F} \tag{7.20}$$

Then, for $m(t) = A_m \sin 2\pi f_m t$, Eq. (7.19) can be simplified to

$$(SNR)_{out,FM} = 3(\beta_f + 1)\beta_f^2 (SNR)_{in,FM} \tag{7.21}$$

At the same time,

$$(SNR)_{out,FM} = 3\beta_f^2 (SNR)_{in,AM} \tag{7.22}$$

Expressions (7.18) and (7.21) to (7.22) are valid only if $(SNR)_{in}$ exceeds the threshold of the FM detector. The minimum received value of $(SNR)_{in}$ needed to exceed the threshold is around 10 dB [3]. Below this threshold, the demod-ulated signal becomes noisy. Equation (7.21) shows that SNR at the output of the FM detector can be increased with increase of the modulation index β_f of the transmitted signal. At the same time, the increase in modulation index β_f leads to an increased bandwidth and spectral occupancy. In fact, for small values of β_f, Carson's rule gives the channel bandwidth of $2\beta_f f_m$. As also follows from (7.21), the SNR at the output of the FM detector is $(\beta_f + 1)$ times greater than the input SNR for an AM signal with the same RF bandwidth. Moreover, it follows from (7.21) that $(SNR)_{out, FM}$ for FM is much greater than $(SNR)_{out, AM}$ for AM.

Finally, we should notice that, as follows from (7.21), the term $(SNR)_{out, FM}$ increases as a cube of the bandwidth of the message. This clearly illustrates why an FM offers very good performance for fast fading signals when compared with an AM. As long as $(SNR)_{in,FM}$ remains above threshold, $(SNR)_{out, FM}$ is much greater than $(SNR)_{in, FM}$. A technique called *threshold extension* is usually used in FM demodulators to improve detection sensitivity to about $(SNR)_{in, FM} = 6$ dB [5].

7.2   Digital Signal Modulation

As was mentioned earlier, *modulation* is the process where the baseband mes-sage information is added to the bandpass carrier. In *digital modulation* the digital beam stream is transmitted as a *message*, and then converted into the analog signal of the type described by (7.1) that modulates the digital bit stream into a carrier signal. As was mentioned above, the analog signal described by (7.1) has amplitude, frequency, and phase. By changing these three character-istics, we can formulate three kinds of digital modulation [3–10]:

Amplitude shift keying (ASK) for phase and frequency keeping being constant
Frequency shift keying (FSK) for amplitude and phase keeping being constant
Phase shift keying (PSK) for amplitude and frequency keeping being constant.

In the so-called hybrid modulation methods combinations of these three kinds of modulation are usually used. When frequency is constant, but amplitude and phase are not constant, a queering (quadrature) amplitude modulation (QAM) occurs. Some modulation methods are linear, such as binary phase shift keying (BPSK), and quadrature phase shift keying (QPSK), including $\pi/4$-QPSK, DQPSK and $\pi/4$-DQPSK. At the same time, FSK as well as minimum shift keying (MSK) and Gaussian minimum shift keying (GMSK) are nonlinear modulation techniques [3–10].

Because digital modulation offers many advantages over analog modulation, it is often used in modern wireless communication systems. Some advantages include greater noise immunity and robustness to channel impairments, easier multiplexing of various forms of information (such as voice, data, and video), and greater security. Moreover, digital transmissions accommodate digital error-control codes, which detect and correct transmission errors, and support complex signal processing techniques such as coding and encryption (see Chapters 4 and 5).

7.2.1 Main Characteristics of Digital Modulation

7.2.1.1 Power Efficiency and Bandwidth Efficiency
The efficiency of each digital modulation technique depends on many factors. The main goal of such a modulation is to obtain in various multipath and fading conditions low bit error rate (BER) at low received SNRs, minimum bandwidth occupation, and so on. The performance of a modulation scheme is often measured in terms of its *power efficiency* and *bandwidth efficiency*. The first term describes the ability of a modulation technique to preserve the fidelity of the digital message at low power levels. In a digital communication system, in order to increase noise immunity, it is necessary to increase the signal power. However, the amount by which the signal power should be increased to obtain a certain level of fidelity, i.e. an acceptable bit error probability, depends on the particular type of modulation employed.

The power efficiency, η_p, of a digital modulation scheme is a measure of how favorably this trade-off between fidelity and signal power is made, and is often expressed as a *ratio of the signal energy per bit to noise PSD*, E_b/N_0, required at the receiver input for a certain probability of error (say 10^{-5}).

The bandwidth efficiency, η_B, describes the ability of a modulation scheme to accomodate data within a limited bandwidth. In general, increasing the data rate implies decreasing the pulse width of a digital symbol, which increases the bandwidth of the signal. Bandwidth efficiency reflects how efficiently the allocated bandwidth is utilized and is defined as the ratio of the *throughput data rate* (bits per second or bps) and a given bandwidth (in hertz), that is, the bandwidth or *spectral* efficiency is measured in bps/Hz. If R is the data rate in bps, and B is the bandwidth (in Hz) occupied by the modulated carrier signal, then bandwidth efficiency is expressed as

$$\eta_B = \frac{R}{B} \text{bps/Hz} \tag{7.23}$$

The system capacity of a digital communication system is directly related to the bandwidth efficiency of the modulation scheme, since modulation with a greater value of η_B will transmit more data in a given spectrum allocation.

However, there is a fundamental upper bound on achievable bandwidth efficiency, which can be defined from well-known Shannon–Hartley channel coding theorem [6]. This theorem states that for an arbitrary probability of error, the maximum possible bandwidth efficiency is limited by the white or Gaussian noise in the channel, which is described by probability density function (PDF) and cumulative distribution function (CDF), presented in Chapter 6, and is given by the channel capacity formula:

$$\eta_{B\max} = \frac{C}{B} = \log_2\left(1 + \frac{S}{N}\right) = \log_2\left(1 + \frac{S}{N_0 B_\omega}\right) \tag{7.24}$$

where C is the channel capacity (in bps), B_ω is the channel bandwidth, N_0 is the signal PSD (in W/Hz), and $\frac{S}{N}$ is the SNR.

7.2.1.2 Bandwidth and Power Spectral Density of Digital Signals

The definition of *signal bandwidth* varies with context, and there is no single definition that covers all applications [6]. All definitions, however, are based on some measure of the PSD of the signal. The PSD of a random signal $x(t)$ is defined as [7]

$$P_x(t) = \lim_{T \to \infty} \left(\frac{\langle |X_T(f)|^2 \rangle}{T} \right) \tag{7.25}$$

where $\langle \cdot \rangle$ denotes an ensemble average, and $X_T(f)$ is the Fourier transform of $x_T(t)$, which is the truncated version of the signal $x(t)$, defined as

$$x_T(t) = \begin{cases} x(t) & -T/2 < t < T/2 \\ 0 & t \text{ elsewhere} \end{cases} \tag{7.26}$$

The PSD of a modulated (bandpass) signal is related to the PSD of its baseband complex envelope. A bandpass signal $s(t)$ is represented as

$$s(t) = \text{Re}[g(t)\exp(j2\pi f_c t)] \tag{7.27}$$

where $g(t)$ is the complex baseband envelope. Then the PSD of the bandpass signal can be presented in a more simplified form than in (7.5), that is,

$$P_s(f) = \frac{1}{4}[P_g(f + f_c) + P_g(-f - f_c)] \tag{7.5a}$$

where $P_g(f)$ is the PSD of $g(t)$. Comparing (7.5) and (7.5a) we can state that the power spectra density of the analog and discrete baseband signal has the same form of presentation in the frequency domain.

The *absolute bandwidth* of a signal is defined as the range of frequencies over which the signal has nonzero PSD. For symbols represented as rectangular baseband pulses, the PSD has a $(\sin f)^2/f^2$ profile that extends over an infinite range of frequencies, and has an absolute bandwidth of infinity. A simpler and more widely accepted measure of bandwidth is the first *null-to-null bandwidth*. The null-to-null bandwidth is equal to the width of the main spectral lobe. A very often used measure of bandwidth that measures the dispersion of the power spectrum is the *half-power bandwidth*, which is defined as the interval between frequencies at which the PSD has dropped to half power, or 3 dB below the peak value. Half-power bandwidth is also called the 3 *dB bandwidth*. According to the definition adopted by the Federal Communication Committee (see [3–9]), the occupied bandwidth is the band that comprises 99% of the signal power: 0.5% of the signal power is above the upper band limit and 0.5% of the signal power below the lower band limit.

7.2.2 Linear Digital Modulation

We present now a few examples of such kind of modulation, transporting the reader to excellent books [4–10] (see also Chapter 4).

Linear modulation is the modulation where the amplitude of the transmitted signal varies linearly with the modulating digital signal $m(t)$ according to the following law:

$$s(t) = \text{Re}[A\, m(t) \exp(j2\pi f_c t)]$$
$$= A[m_R(t) \cos(2\pi f_c t) - m_I(t) \sin(2\pi f_c t)]$$
$$m(t) = m_R(t) + jm_I(t) \tag{7.28}$$

This kind of modulation has a good spectral efficiency, but linear amplifiers have *poor* power efficiency [3–10]. Side lobes are generated, increasing adjacent channel interference, and cancel the benefits of linear modulation.

7.2.2.1 Amplitude Shift Keying (ASK) Modulation

This is a modulation where the carrier sinusoid is keyed (or switched) *on* if the input bit is "1" and *off* if "0" (the so-called on–off keying [OOK] [3, 5, 10]). This kind of modulation is shown in Figure 7.3.

Figure 7.3 The message $m(t)$ unipolar signal (a) and the baseband modulated OOK signal (b).

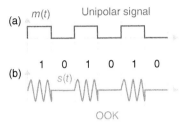

7.2.2.2 Binary Phase Shift Keying (BPSK) Modulation

BPSK modulated signals $g_1(t)$ and $g_2(t)$ are given by

$$g_1(t) = \sqrt{\frac{2W_b}{T_b}}\cos(2\pi f_c t), \quad 0 \le t \le T_b \tag{7.29a}$$

and

$$g_2(t) = -\sqrt{\frac{2W_b}{T_b}}\cos(2\pi f_c t), \quad 0 \le t \le T_b \tag{7.29b}$$

where W_b is the energy per bit, T_b is the bit period, and a rectangular pulse shape $p(t) = \Pi((t - T_b/2)/T_b)$ is assumed. Basis signals ϕ_i for this signal, set in 2D vector space, simply contain a single waveform ϕ_1, where

$$\phi_1(t) = \sqrt{\frac{2}{T_b}}\cos(2\pi f_c t), \quad 0 \le t \le T \tag{7.30}$$

The result of such modulation is presented in Figure 7.4.

Using this basis signal, the BPSK signal set can be represented as

$$g_{iBPSK} = \left\{ \sqrt{W_b}\phi_1(t), -\sqrt{W_b}\phi_1(t) \right\} \tag{7.31}$$

Such a mathematical representation of a vector, consisting of two points that are then placed at the *constellation diagram*, is shown in Figure 7.5, which provides a graphical representation of the complex envelope of each possible symbol state. The distance between signals on a constellation diagram relates to how different the modulation waveforms are, and how well a receiver can differentiate between all possible symbols when random noise is present.

As was mentioned in Chapter 4, the number of basis signals will always be less than or equal to the number of signals in the set. The number of basis signals required to represent the complete modulation signal set is called the *dimension* of the vector space (in the example above – it is two-dimensional [2D]

Figure 7.4 BPSK signal presentation.

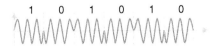

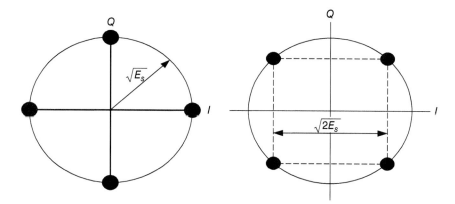

Figure 7.5 Constellation diagram of QPSK and $\pi/4$-QPSK modulated signals.

vector space). If there are many basis signals in the modulation signal set, then all of them must be orthogonal according to (7.29a) and (7.29b) with sign "+" and "−", respectively.

7.2.2.3 Quadrature Phase Shift Keying (QPSK) Modulation

The advantage of the QPSK signal is that it has twice the bandwidth efficiency or two bits at a time:

$$s_{QPSK}(t) = \sqrt{\frac{2E_s}{T_s}}\cos\left[2\pi f_c t + i\frac{\pi}{2}\right] \quad 0 \leq t \leq T_s \quad i = 0, 1, 2, 3$$

$$= \sqrt{\frac{2E_s}{T_s}}\cos\left(i\frac{\pi}{2}\right)\cos(2\pi f_c t) - \sqrt{\frac{2E_s}{T_s}}\sin\left(i\frac{\pi}{2}\right)\sin(2\pi f_c t)$$

$$(7.32)$$

This signal set is shown geometrically in Figure 7.5, where the left diagram is for pure QPSK and the right one is for $\pi/4$-QPSK modulation, which is with angle shift at $\pi/4$.

7.2.3 Nonlinear Digital Modulation

As was mentioned at the beginning of this section, the FSK signals are examples of nonlinear type of digital modulation. We briefly describe them.

7.2.3.1 Frequency Shift Keying (FSK) Modulation

In FSK modulated signals, switching the carrier sinusoid frequency f_c to $f_c - \Delta f$ occurs if the input bit is "0" and to $f_c + \Delta f$ if the input bit is "1." The results of modulation are shown in Figure 7.6.

Figure 7.6 FSK modulated signal presentation.

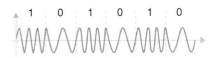

Concluding this chapter, we should notice that usage of each kind of modulation depends on the conditions of propagation inside a channel, effects of fading inside, and on what kinds of detectors and the corresponding filters are used. All these aspects are fully described in excellent books [3–5, 7–10].

Problems

7.1 A zero mean ($\langle m(t)\rangle = 0$) sinusoidal message is applied to a transmitter that radiates the AM signal with a power of 10 kW. The modulation index $k_m = 0.6$.
  Find: The carrier power. What is it in percentage of the total power in the carrier? What is the power of each sideband?

7.2 A sinusoidal modulating signal, $m(t) = 4\cos 2\pi f_m t$, with $f_m = 4\,\text{kHz}$ and the maximum amplitude $A_m = 4\,\text{V}$, is applied to an FM modulator, which has a frequency deviation constant gain $k_f = 10\,\text{kHz/V}$.
  Find: The peak frequency deviation, Δf; the modulation index, β_f.

7.3 A frequency modulated signal, having carrier frequency $f_c = 880\,\text{MHz}$ and the sinusoidal modulating waveform with frequency $f_m = 100\,\text{kHz}$, has a peak deviation of $\Delta f = 500\,\text{kHz}$.
  Find: The receiver bandwidth necessary to pass through the detector of such a signal.

7.4 An FM signal with $f_m = 5\,\text{kHz}$ and with a modulation index $\beta_f = 3$.
  Find: The bandwidth required for such an analog FM. How much output SNR improvement would be obtained if the modulation index will be increased to $\beta_f = 5$. What is the trade-off bandwidth of this improvement?

References

1 Jakes, W.C. (1974). *Microwave Mobile Communications*. New Jersey: IEEE Press.
2 Steele, R. (1992). *Mobile Radio Communication*. IEEE Press.
3 Rappaport, T.S. (1996). *Wireless Communications*. New York: Prentice Hall PTR.

4 Stuber, G.L. (1996). *Principles of Mobile Communication*. Boston-London: Kluwer Academic Publishers.

5 Couch, L.W. (1993). *Digital and Analog Communication Systems*. New York: Macmillan.

6 Lusignan, B.B. (1978). Single-sideband transmission for land mobile radio. *IEEE Spectrum*: 33–37.

7 Ziemer, R.E. and Peterson, R.L. (1992). *Introduction to Digital Communications*. Macmillan Publishing Co.

8 Saunders, S.R. (1999). *Antennas and Propagation for Wireless Communication Systems*. New York: Wiley.

9 Proakis, J.G. (1989). *Digital Communications*. New York: McGraw-Hill.

10 Krouk, E. and Semenov, S. (eds.) (2011). *Modulation and Coding Techniques in Wireless Communications*. Chichester, England: Wiley.

8

Optical Sources and Detectors

8.1 Emission and Absorption of Optical Waves

The processes due to which light is emitted and absorbed by atoms should be described in terms of the corpuscular (or particulate) nature of light accounting for the wave–particle dualism well known from the principles of quantum theory [1–14]. Because explanation of the main aspects of quantum theory is not the aim of this book, we briefly introduce the reader to the physical aspects of light wave sources based on the primitive description of the subject of this chapter from the regular school physics fundamentals.

As follows from classical physics, the atom was held to possess natural resonance frequencies. These corresponded to the electromagnetic wave frequencies that the atom was able to emit when excited into oscillation. Conversely, when light radiation at any of these frequencies fell upon the atom, the atom was able to absorb energy from the radiation in the way of all classical resonant system–driving force interactions.

At the same time, the theory leading from the classical theory cannot explain some features observed – for example, in a gas discharge, where some frequencies that are emitted by the gas are not also absorbed by it under quiescent conditions, or during observation of photoelectric effects, where electrons are ejected from atoms by the interaction with light radiation. It was found experimentally and stated theoretically that the energy with which electrons are ejected depends not on the intensity of the light but *only on its frequency*.

The explanation of these facts follows from quantum theory, according to which atoms and molecules can exist *only in discrete energy levels*. These energy levels can be listed in order of ascending discrete values of energy: E_0, E_1, E_2, ..., E_n. Under conditions of thermal equilibrium the number of atoms having energy E_i is related to the number of atoms having energy E_j by the Boltzmann relation [2, 7, 8]

$$\frac{E_i}{E_j} = \exp\left\{-\frac{E_i - E_j}{kT}\right\} \tag{8.1}$$

Fiber Optic and Atmospheric Optical Communication, First Edition.
Nathan Blaunstein, Shlomo Engelberg, Evgenii Krouk, and Mikhail Sergeev.
© 2020 John Wiley & Sons, Inc. Published 2020 by John Wiley & Sons, Inc.

where $k = 1.38 \cdot 10^{-23}$ J/K is Boltzmann's constant and T is the absolute temperature (in kelvin, K). According to quantum theory, and the following "wave–particle" dualism, light waves can be presented as a discrete stream of light particles called *photons* (or *one quantum of light*) with the individual energy for each photon of number i, depending only on its own frequency v_i, that is, $E_i = hv_i$. Here, h is Planck's quantum constant, which equals $6.625 \cdot 10^{-34}$ J · s. Therefore, each material, as the atomic system consisting of "i–j" steady states of atoms and electrons distribution inside this system, can absorb only when its frequency v corresponds to at least one of the values v_{ji}, where

$$hv_{ji} = E_j - E_i, \quad j > i \tag{8.2}$$

We use the same symbol for the frequency, as is usually used in quantum electrodynamics, to differentiate it with the symbol $f = \omega/2\pi$ usually used in classical electrodynamics. In this case, Eq. (8.2) states that one quantum of light, (called *photon*) with energy hv_{ji} can be absorbed by the atom, which in consequence has increased in energy from one of its steady states of the atomic system with energy E_i to another steady state of the atomic system with energy E_j. Correspondingly, a photon will be emitted when a downward transition occurs from E_j to E_i, and this photon will have the same frequency v_{ji}.

In this context, we can present the light radiation (presented in Chapter 2 in view of electromagnetic wave) as a stream of photons. If there is a flux of q photons across unit area per unit time, then we can write for the light intensity the following expression:

$$I = qhv \tag{8.3}$$

Similarly, any other quantity defined within the wave context also has its counterpart in the corpuscular context. So, in our further explanation of matter, we will use both the wave and the corpuscular (e.g. particle) representation.

Thus, each atom or molecule of any material has a characteristic set of energy levels, called *steady-state conditions* of the atoms, and free electrons inside the material, as an atomic system. If so, the light frequencies v_{ji} emitted in the form of photons or absorbed from photons by atoms and molecules fully characterize each material under consideration. When an excited system returns to its lowest state, some return pathways are more probable than others, and these probabilities are also characteristic of the particular atoms or molecules under consideration.

In other words, the light waves, as electromagnetic continuous waves, can be regarded as a probability function whose intensity at any point in space defines the probability of finding a photon there. According to this *wave–particle dualism*, the emission and/or the absorption spectrum of any material can be used for its identification and to determine the quantity present. These ideas form the substance of the subject known as spectroscopy, which is a very extensive and powerful tool in material analysis, but outside the scope of our book.

8.2 Operational Characteristics of Laser

The laser is a very special source of light, the discovery of which in 1960 by T. H. Maiman [1] gave a push to optical fiber and wireless communication. The word *laser* is an acronym for *light amplification by stimulated emission of radiation*, and we will describe briefly the processes on which it depends.

As was mentioned in the previous section, the photon could cause an atomic system to change from one of its steady states to another according to the process described by (8.2), that is, the change of the system from a lower to a higher energy state. However, if the system was already in the higher of the two states when the photon acted, then this action will cause a transition down to the lower state, still in accordance with (8.2), but now with $j < i$. This process is called *stimulated emission* since the effect is to cause the system to emit a photon with energy $h\nu_{ji}$ corresponding to the energy lost by the system. Finally, we have two kinds of photons – the *acting photon* and the *emitted photon*. This process is crucial to laser action.

We must also mention the fact that a system that is not in its lowest energy state is not in stable equilibrium. If it has any interaction with the outside background, it will eventually fall to its lowest state. Thus, an atomic system in a state E_i will fall spontaneously to the lower state E_j even without the stimulus of $h\nu_{ji}$ in a time, which depends on the exact nature of equilibrium conditions. The photon that results from this type of transition is thus said to be due to *spontaneous emission*.

To understand how laser works, let us consider a two-level atomic system with energy level E_0 and E_1 (see Figure 8.1a).

Suppose that we illuminate this system with electromagnetic radiation of frequency $\nu_{10} = (E_1 - E_0)/h$. Initially, if the system is in thermal equilibrium at temperature T, the relative numbers of atoms in the two levels will be, according to formula (8.1),

$$\frac{N_1}{N_0} = \exp\left\{ -\frac{E_1 - E_0}{kT} \right\} \tag{8.4}$$

As follows from (8.4), if $E_1 > E_0$ then $N_1 < N_0$. Suppose now that the intensity of the radiation at frequency ν_{10} is steadily increased from zero. At low levels, if it is assumed that the probability of transition is the same for the two transition directions, more atoms will be raised from the lower to the higher state than vice versa since, according to (8.4), there are more atoms in the lower state. As the intensity is increased, the number of downward transitions (stimulated and/or spontaneous) will increase as the occupancy of the upper state

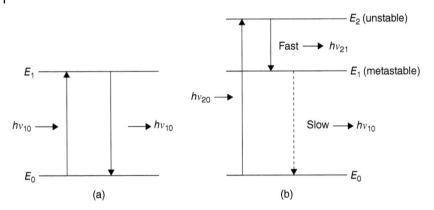

Figure 8.1 Elementary processes of emission and absorption of light photons.

rises, tending toward the saturation condition where the occupancies of the two states and the rates of transition in the two directions are equal.

Considering now a three-level system shown in Figure 8.1b, we have a lowest level E_0, a metastable level E_1, and an unstable level E_2. If this system, being initially in thermal equilibrium, is irradiated with light frequency $v_{20} = (E_2 - E_0)/h$, the effect is to raise a large number of atoms from E_0 to E_2. These then decay quickly to the state E_1 by spontaneous emission only (since the input light frequency does not correspond to this transition), and subsequently only slowly from this metastable (i.e. long-lived) state back to the ground state. Owing to this process, a larger number of atoms can be in state E_1 than in state E_0. Since this does not correspond to Boltzmann distribution, it is known as an inverted population [2].

Suppose that a second beam of light is incident on this inverted population at frequency $v_{10} = (E_1 - E_0)/h$. This light corresponds to a situation where it can more frequently produce downward transitions by stimulated emission from E_1 to E_0 than it can excite atoms from E_0 to E_1. Thus, more stimulated photons are produced than are absorbed by excitation, and this beam receives *gain* from the material (or media), i.e. it is amplified. The medium is said to be *pumped* by the first beam to provide gain for the second. We have *light amplification by stimulated emission of radiation*, that is, we have obtained *laser* [3–12].

If now the medium is enclosed in a tube with parallel mirrors at the ends, as shown in Figure 8.2, the stimulated photons can move between the mirrors and they act to stimulate even more photons. So, we have proved the *amplifier* with positive feedback and have produced an *oscillator*.

If one of the two mirrors is only partially reflecting, some of the oscillator energy can emerge from the tube. This energy will be in the form of a light wave of frequency $v_{10} = (E_1 - E_0)/h$, which is accurately defined if the energy levels are sharp. As for relatively large intensity, if the volume of the tube is

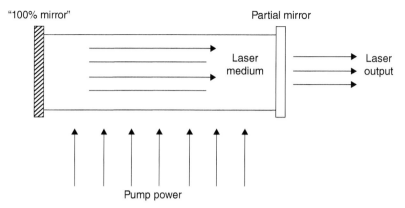

Figure 8.2 Simple scheme of gaseous laser action.

large, the pump power is large, and the cross-sectional area of the tube is small and well collimated since the light will only receive amplification within the tube, if it is able to *oscillate* between the two parallel mirrors, and has an accurately defined phase (the stimulated photon is equal to that which stimulated it). Finally, we have monochromatic (with narrow frequency band), coherent (with well-defined phase), and well collimated light: we have *laser light*.

As mentioned above, light interacts with matter as a result of *emission* or *absorption* of photons by the individual atoms or molecules. Considering light incident upon matter, we assumed that the incident light consists of a stream of photons that are guided by electromagnetic wave. However, this process is random and a stream of individual particles cannot have an arrival rate that is constant in time.

Let us begin with the assumption that atoms in excited states emit photons in a random manner when falling spontaneously to lower states, i.e. nobody knows which particular atoms will emit in any given time interval. As was noted above, for moderate light intensities a very small fraction of the total number of atoms in an emitting material will effect transitions in the desired detection times.

Let us suppose the existence of an assembly of n atoms, and the probability of any one of them emitting a photon in time τ is p. Then, the average number of photons detected in this time would be np, but the actual number for r photons will vary statistically around this mean according to Poisson law with probability

$$P_r = \exp(-np)\frac{(np)^r}{r!} \tag{8.5}$$

Thus, the probability of receiving zero photons is $\exp(-np)$, and two photons is $\exp(-np) \cdot (np)^2/2!$, and so on.

We can relate np to the mean optical power received by the detector, P_m, for np is just the mean number of photons received in time τ. Hence

$$P_m = np\frac{h\nu}{\tau} \tag{8.6}$$

and then the mean of the distribution becomes [9]

$$np = \frac{P_m\tau}{h\nu} = \frac{P_m}{h\nu B_\omega} \tag{8.7}$$

where B_ω is the detector bandwidth. We need to measure the spread from this mean, which is called the variance σ_n^2. According to Poisson law, this variance is equal to the mean, that is, $\sigma_n^2 = np$. Finally, we obtain the variance as a measure that gives us the "noise" of the signal. This noise is called the *quantum noise* or the *photon noise*, the power of which equals

$$N \equiv \sigma_n = \left(\frac{P_m}{h\nu B_\omega}\right)^{1/2} \tag{8.8}$$

Consequently the signal-to-noise ratio (SNR) will be [9]

$$\text{SNR} = \frac{P_m}{B_\omega h\nu} \cdot \frac{1}{N} = \left(\frac{P_m}{h\nu B_\omega}\right)^{1/2} \tag{8.9}$$

This is an important result since it provides the ultimate limit on the accuracy with which a light power level can be measured. We note that the measurement accuracy improves as $(P_m)^{1/2}$, and for lower power, the accuracy will be poor enough. Correspondingly, if frequency ν is larger for a given power, the accuracy is worse. Finally, it must be remembered that these conclusions only apply when the probability of photon emission is very small. This assumption is no longer valid for intense laser beams of power density $W \geq 10^6$ W/m². Such light is sometimes said to be non-Poisson statistics [9].

8.3 Light-Emitting Sources and Detectors

The most commonly used light sources in optical communication are the light-emitting diode (LED) and the laser diode (LD) [6–15].

8.3.1 Light-Emitting p–n Type Diode

The LEDs are semiconductors with positive–negative (p–n) junction. In this LED, a junction is constructed from a p-type to n-type semiconductor, separated by a depletion zone with an energy (see Figure 8.3a). The power distribution inside the semiconductor when a source with voltage is introduced is shown in Figure 8.3b.

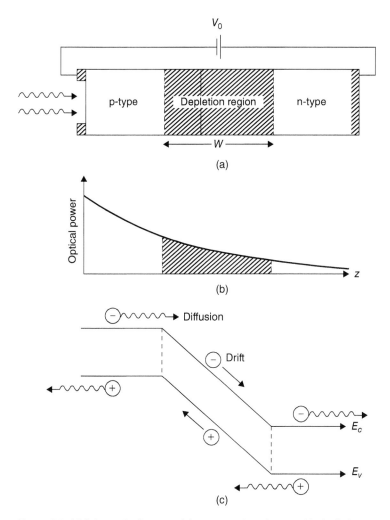

Figure 8.3 (a) Schematic diagram of the p–n semiconductor with depletion zone of energy W, (b) distribution of optical power along a distance inside a depletion zone, and (c) schematically presented drift and diffusion of electron (−) and holes (+) through the depletion zone.

The p–n junction comes to equilibrium by diffusion of majority carriers (*holes* (+) and *electrons* (−), respectively) across the physical junction (i.e. depletion zone, see Figure 8.3c) until an equilibrium is established between the force exerted by the field (due polarization of charge) and the tendency to diffuse. If an external electric field (with voltage V_0, see Figure 8.3a) is imposed on the junction (i.e. depletion zone) in opposition to the equilibrium field, the result is to cause electrons and holes with differing energy levels to annihilate each other and thus give rise to photons equal in energy to the bandgap of the material.

Thus, in LED free charges (electrons and holes) injected into the junction region chaotically and spontaneously recombine with the subsequent emission of radiation. The two best known from those semiconductors are gallium-arsenide (GaAs) semiconductor with the depletion zone (called *bandgap*) energy of $W_g = 1.4$ eV, operating at wavelength of 0.89 μm, and gallium-indium-phosphate (GaInP) semiconductor with the depletion energy of $W_g = 1.82$–1.94 eV, operating at 0.64–0.68 μm. More information on different kinds of light sources can be found in Refs. [4–9, 14].

Ideally, the output optical power increases linearly with input current. Thus, the light power P is a replica of the driving current I. In the forward-biased region, the output power is given by

$$P = A_i I \tag{8.10}$$

The output wavelength is determined by the bandgap energy, W_g, of the semiconductor material. In particular, the output wavelength is given by [14]

$$\lambda = \frac{1.24}{W_g} \tag{8.11}$$

In formula (8.11), the wavelength is in micrometers (μm) and the bandgap energy is in electron volts (eV), where $1\,\text{eV} = 1.6 \cdot 10^{-19}$ J.

An important property of light sources based on LED and used in fiber optic communications is its *spectral width*. Simply speaking, this characteristic defines the range of wavelengths (or frequencies) over which it emits significant amount of power. More exactly, this characteristic was defined in [5], according to which the diode 3 dB decay bandwidth is defined by the average of the transit times for electrons and holes during their diffusion process, shown in Figure 8.3c; that is,

$$f|_{3dB} = \frac{0.44}{\tau} \tag{8.12}$$

Ideally, a light source based on LED would be monochromatic, emitting a single wavelength. In practice, no such light source exists. All of them radiate over a range. For LEDs, the range is typically on the order of 20–100 nm. Coherence of such sources refers to how close the source radiation is to the ideal single wavelength:

The smaller the spectral width, the more coherent the source.

8.3.2 Laser p–n Type Diode

The laser diode (LD) has a number of similar characteristics as the LED. It is also a p–n junction semiconductor that emits light when forward biased.

Light amplification occurs when photons stimulate free charges in the junction region to recombine and emit. The light beam is reflected back and forth through the amplifying medium (let say, gas) by reflectors at each end of the junction. The amplification together with the feedback produces an oscillator emitting at optical frequencies.

As for the laser diodes (LDs), the stimulated recombination leads to radiation that is more coherent than that produced by spontaneous recombination occurring in LED. Therefore, the spectral widths for LDs are typically in the range of 1–5 nm, even less.

We should notice that in LDs the output power does not increase until the input current is beyond a threshold value (I_{Th}). Thresholds are on the order of a few to a few tens of mA. Voltages are on the order of a few volts (V) and output powers of a few mW. The materials used for LDs are the same as those used in constructing LEDs. We do not enter into a description of the technology of design of both kinds of diodes because this is out of the scope of this book (as was mentioned from the beginning in the chapter); the reader is referred to the corresponding books [3–9, 15].

8.3.3 Photodiode

The processes that enable light powers to be measured accurately depend directly upon the interaction between photons and matter. In most cases of quantitative measurement, the processes rely on the photon to raise an electron in an atom to a state where it can be observed directly as an electric current.

Let us consider again the p–n junction presented in Figure 8.3. When considering the process mentioned above, it was noted that the physical contact between these two types of semiconductor (i.e. p and n) led to a diffusion of majority carriers across the junction in an attempt to equalize their concentration on either side. The result, however, was to establish an electric field across the junction as a consequence of the charge polarization.

Suppose now that a photon is incident upon the region of the semiconductor exposed to this field. If this photon has sufficient energy to create an electron–hole pair, these two new charge carriers will be swept quickly in opposite directions across the junction to give rise to an electron current that can then be measured. The process is assisted by application of an external "reverse bias" electric field. This simple picture of the process enables us to establish *two important relationships* appropriate to such devices, called *photodiodes* (PD).

First, for the photon to yield an electron–hole pair its energy must satisfy $h\nu > W_g$, where, again, W_g is the bandgap energy of the material of the semiconductor. If frequency ν of the photon is too high, however, all the photons will be absorbed in a thin surface layer and the charge pairs will not be collected efficiently by the junction. Thus, there is a frequency "responsivity" spectrum for each type of photodiode, which, consequently, must be matched to the spectrum of the light which is to be detected.

Secondly, suppose that we are seeking to detect a light power of P at an optical frequency v. This means that P/hv photons are arriving per second. Suppose now that a fraction η of these photons produces electron–hole pairs. Then, the charge carriers of each sign (electron and hole) produced every second are proportional to $\eta P/hv$. If so, all the collected electrons and holes give rise to the observed electric current, given by

$$I = e\eta P/hv \tag{8.13}$$

Thus, the current is proportional to the optical power. This means that the electrical power is proportional to the square of the optical power. It is important, therefore, when satisfying the SNR for a detection process to be sure about whether the ratio is stated in terms of electrical and optical power.

8.3.4 PiN and p–n Photodiodes – Principle of Operation

Light intensity direct detection (DD) and intensity modulation (IM) are the basic principles of most present-day fiber optic sensing and data transmission systems. A laser diode (LD) and LED described above are sources that operate with intensity-modulated signals, by modulating the drive current I and a photocurrent I_{ph}, which is directly proportional to the received power generated at the detector. Moreover, the mode of operation of the p–n and positive-intrinsic-negative (PiN) diodes has a little difference in basic principles, which depends on their structure that differs mostly on the width of depletion rejoin required. Therefore in our further explanation of different kinds of optical receivers, we will not separate their operational characteristics that are based on similar basic physical parameters of both kinds of diodes.

The detector considered is the simplest form of semiconductor detector, i.e. the PiN diode. In this diode, there is an intrinsic layer between the p-region and the n-region as shown in Figure 8.4. Therefore, this kind of photodiode is called a PiN diode. The intrinsic layer is used to increase the absorption of photons, i.e. the optical power. The size of this layer is a trade-off between sensitivity and response time of the diode.

The *DD* process of such a diode, a scheme of which is shown in Figure 8.3a, is illustrated by Figure 8.3c. Photons with energy greater than the bandgap energy $W_g \equiv W$ enter the semiconductor and are able to excite electrons from the valence band to the conduction band.

If the electrons and holes produced reach the junction or are created within the depletion region, they can be swept through the device and so generate a photocurrent. The depth over which optical absorption occurs is characterized through the absorption coefficient $a = a(hv)$. If $\Phi(x)$ is the optical intensity

Figure 8.4 Scheme of PiN detector (a) and spatial distribution of PiN's inner electric field (b).

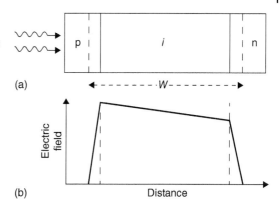

within the material at the depth x,

$$\Phi(x) = \Phi(0) \exp\{-\alpha x\} \qquad (8.14)$$

$\alpha = \alpha(v)$ is important because it is a measure of the thickness of material required to absorb the radiation. For example, if $x = 2/\alpha$, 86% absorption is achieved, and if $x = 3/\alpha$ this rises to 95% [4–9, 15].

The important parameter of the direct detector presented in Figure 8.3a is the quantum efficiency η introduced by formula (8.12). We define this parameter more precisely as a ratio of the number of photogenerated electrons, which are collected (i.e. which transverse the depletion region [see Figure 8.3c] and contribute to the photocurrent) to the number of photons that are incident on the detector. The definition strictly refers to the external quantum efficiency of the device, which takes into account losses due to reflection at the detector surface. If, as in (8.13), Φ is the photon intensity and J_{ph} is the photocurrent density, then

$$\eta = \frac{J_{ph}}{e\Phi} \qquad (8.15)$$

where e is the charge of the electron. Denoting the optical power incident on the detector surface of area S as P_r, we can rewrite (8.15) as

$$\eta = \frac{I_{ph} h v}{e P_r} \qquad (8.16)$$

This leads to the definition of direct detector responsivity R [9–14] as

$$R = \frac{e \eta}{h v} \; (\text{A/W}) \qquad (8.17)$$

Finally, the direct detector photocurrent $I_{ph} = J_{ph} \cdot S$ is given by

$$I_{ph} = R \cdot P_r \qquad (8.18)$$

Therefore, the optical signal with power $P_S(t)$ and frequency f_S, passing via PiN detector, transforms into an electrical current $I_S = R \cdot P_S(t)$.

A high value of external quantum efficiency η in such kinds of photodetectors depends on the following:

- Reducing reflections from the detector surface, achieved by use of an antireflection coating
- Maximizing absorption within the depletion region, which depends on device design and requires $W \approx 2/\alpha \div 3/\alpha$ (see Figure 8.4)
- Avoiding carrier recombination, achieved through device design based on minimization of photon absorption outside the depletion region.

We should notice that for PiN diode detector reverse bias is normally applied so that a wide depletion zone is created and carrier generation predominantly takes place there. Carriers are swept through by the drift field with little or no recombination. Generation of electron–hole pairs outside the depletion region relies upon the process of diffusion to drive carriers toward the junction and hence contribute to the photocurrent. In the event that photogenerated electrons and holes recombine before reaching the junction they do not contribute. Hence, carrier generation outside the depletion zone can lead to recombination losses and addition affects the rise and fall time of the detector, thus influencing the speed and bandwidth.

As was shown in Ref. [9], the role of a junction that determines the depth of the depletion zone can be characterized by the detector capacitance C_D that determines the electrical and noise parameters of the optical receiver. This capacitance depends on the energy of the depletion width W (see Figures 8.3a and 8.4) and on the material permittivity, that is, $C_D \propto \varepsilon/W$, where $\varepsilon = \varepsilon_r \varepsilon_0$ (see Chapter 2).

Moreover, the width of depletion zone ultimately limits the transit time for electrons (and/or holes) to drift across the depletion zone, and therefore, the frequency band (i.e. response) of the detector, and for a mean transit time $\bar{\tau}$, this response equals at the 3 dB detector level [5]: $f|_{3dB} \propto \bar{\tau}^{-1}$. For most PiN and p–n diodes, the total noise, which influences the forward current occurring inside the diode, depends both on diode current operated in dark conditions, I_D, and on photocurrent of diode operated in light conditions, I_{ph}; that is, the total noise, N_{PiN}, in the PiN detector equals

$$N_{PiN} = 2e[I + 2I_D + I_{ph}] \qquad (8.19)$$

where

$$I + I_D = I_0 \exp\left\{\frac{eV}{nk_B T}\right\} \qquad (8.20)$$

is the diode current in dark conditions. Here, V is the voltage, $k_B = 1.38 \cdot 10^{-223}$ J/K is the Boltzmann constant, and T is the temperature of environment in Celsius.

8.4 Operational Characteristics of Light Diodes

As follows from the above sections, the most common photodetector for optical communications (fiber and wireless) is the semiconductor junction photodiode, which converts optical power to an electric current. Then the received current is [9, 14, 15]

$$I_{ph} = R \cdot P \tag{8.21}$$

where, as above, R is the photodetector responsivity defined by (8.17), and the power at the receiver, P, is defined by a product of the power incident on the detector surface P_r, introduced above, and the photoconductive gain G of the detector, that is,

$$P = P_r \cdot G \tag{8.22}$$

The photodetector materials, their operating wavelengths, and their peak responsivities appear in Table 8.1, selected in Ref. [14] for the design of usually used semiconductors in photodetectors.

The cutoff wavelength is determined by the bandgap energy (e.g. the depletion zone width; see Figure 8.3 or 8.4) and is given by

$$\lambda_c = \frac{1.24}{W_g} \tag{8.23}$$

As in Eq. (8.11) introduced for the emission of an LED and LDs, in Eq. (8.23) the wavelength is in micrometers and the bandgap energy is in eV. It is clear that only wavelengths equal to or smaller than the cutoff wavelength can be detected.

According to Eq. (8.21) the photodetector acts like a constant current source. Therefore, the output voltage $V = I \cdot R_L$ can be increased by increasing the load resistance R_L. However, the receiver bandwidth is no larger than

$$B_\omega = \frac{1}{2\pi R_L C_d} \tag{8.24}$$

so that doing so decreases the receiver bandwidth. The photodiode's shunt capacitance is C_d.

Table 8.1 Main parameters of semiconductors.

Material	λ (nm)	$\lambda(P_{max})$ (nm)	R (A/W)
Si	300–1100	800	0.5
Ge	500–1800	1550	0.7
InGaAs	1000–1700	1700	1.1

Because the noise source is the fundamental characteristic of all kinds of diodes (see above) that characterizes the photoconductive process, we will briefly mention here that there are two kinds of noises inside each kind of photodetector: a thermal noise (called Johnson [4–9]) and generation–recombination noise [9–15]. The Johnson noise spectral density is directly proportional to the absolute temperature T (in kelvin) and inversely proportional to the bulk resistance of the photoconductor, R_S, i.e.

$$N_T = \frac{4 \cdot T}{R_S} \tag{8.25}$$

As for the generation–recombination noise that arises from fluctuations in generation and recombination rates of electron–hole pairs due to the process of photoemission, its spectral density can be presented according to [9, 15] as

$$N_{\text{ph}} = \frac{4e \cdot G \cdot I_{\text{ph}}}{1 + 4\pi^2 \cdot f^2 \cdot \tau_R^2} \tag{8.26}$$

where τ_R is the mean electron–hole recombination time, and f is the 3 dB bandwidth that can be defined as [9, 15]

$$f|_{\text{3dB}} = \frac{1}{2\pi \cdot G \cdot t_T} \tag{8.27}$$

where t_T is the detector transit time.

Concluding this chapter, we should emphasize that all types of photodetection quite readily work up to optical modulation frequencies of about 1 GHz. To exceed this number requires special attention to materials of semiconductors and design, to achieve bandwidths up to 30 GHz and more. For precise and extensive information on light sources and detectors, the reader is referred to the corresponding works [6–15].

References

1 Maiman, T.H. (1960). Stimulated optical radiation in ruby masers. *Nature* 187: 493–503.

2 Yariv, A. (1976). *Introduction in Optical Electronics*, Chapter 5. New York: Holt, Rinehard, and Winston.

3 Kressel, H. and Butler, J.K. (1977). *Semiconductor Lasers and Heterojunction LEDs*. New York: Academic Press.

4 Kressel, H. (ed.) (1980). *Semiconductor Devices for Optical Communications*. New York: Springer-Verlag.

5 Sze, S.M. (1985). *Semiconductor Devices: Physics and Technology*. New York: Wiley.

6 Agrawal, G.P. and Dutta, N.K. (1986). *Long-wavelength Semiconductor Lasers*. New York: Van Nostrand Reinhold.

7 Siegman, A.E. (1986). *Lasers*. University Science Books: Mill Valley, CA.

8 Ebeling, K.J. (1993). *Integrated Optoelectronics*. New York: Springer-Verlag Publishing Co.

9 Dakin, J. and Culshaw, B. (eds.) (1988). *Optical Fiber Sensors: Principles and Components*, vol. 1. Boston-London: Artech House.

10 Coldren, L.A. and Corzine, S.W. (1995). *Diode Lasers and Photonic Integrated Circuits*. New York: Wiley.

11 Marz, R. (1995). *Integrated Optics: Design and Modeling*. Norwood, MA: Artech House.

12 Morthier, G. and Vankwikelberge, P. (1997). *Handbook of Distributed Feedback Laser Diodes*. Norwood, MA: Artech House.

13 Murphy, E.J. (1999). *Integrated Optical Circuits and Components, Design and Applications*. New York: Marcel Dekker.

14 Palais, J.C. (2006). Optical communications. In: *Handbook: Engineering Electromagnetics Applications* (ed. R. Bansal). New York: Taylor & Frances.

15 Dakin, J. and Culshaw, B. (eds.) (1988). *Optical Fiber Sensors: Principles and Components*, vol. 2. Boston-London: Artech House.

Part III

Wired Optical Communication Links

9

Light Waves in Fiber Optic Guiding Structures

In this chapter, we briefly present information on different types of fiber optic structures, following Refs. [1–7].

9.1   Propagation of Light in Fiber Optic Structures

9.1.1   Types of Optical Fibers

The first usually used kind of fiber optic structure is the *step-index fiber* (see Figure 9.1).

The inner structure of each kind of fiber optic is schematically presented in Figure 9.2.

As is clearly seen from Figure 9.2, such kind of fiber consists of a central core of radius a and refractive index n_1, surrounded by a cladding of radius b and refractive index n_2.

According to the definition of TIR (see definition in Chapter 2), to obtain the total reflection from the cladding, its refractive index should be lower than that for the core, i.e. $n_1 > n_2$. Figure 9.3 shows the geometry of optical ray propagation within the core on the assumption that the cladding width is thick enough to exclude the evanescent field decay inside the cladding depth. So, from the beginning we can suppose that the effects of a finite cladding thickness are negligible, and a ray field is small enough to penetrate to the outer edges of the cladding. As will be described below, in multimode *step-index* fiber a large modal distortion occurs.

To avoid such drawbacks of this kind of fiber, a new type called *graded-index* fiber was developed that has the same configuration as the previous fiber, and shown in Figure 9.1. The difference between both kinds of fibers is defined by differences in profiles of the refractive indexes of the core and cladding, as illustrated in Figure 9.1. Thus, as clearly seen from the illustrations, in the *step-index* fiber the index change at the core–cladding interface is abrupt, whereas in the *graded-index* fiber the refractive index decreases gradually inside the core.

Fiber Optic and Atmospheric Optical Communication, First Edition.
Nathan Blaunstein, Shlomo Engelberg, Evgenii Krouk, and Mikhail Sergeev.
© 2020 John Wiley & Sons, Inc. Published 2020 by John Wiley & Sons, Inc.

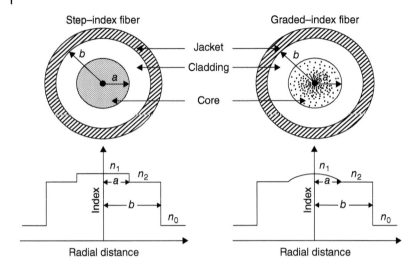

Figure 9.1 Difference between the refractive indexes profiles for step-index and graded-index fibers.

Figure 9.2 A view of the fiber optic inner structure.

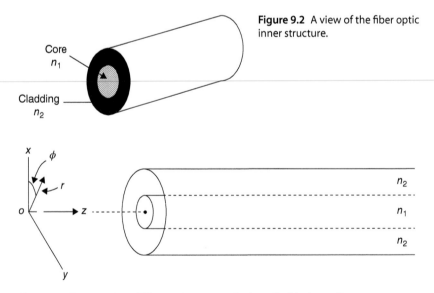

Figure 9.3 Presentation of fiber optic structure in the cylindrical coordinate system.

To understand the effects of wave propagation in both kinds of fibers, let us consider a similar 3D problem of ray propagation in cylindrical waveguide, but now having more complicated geometry by consisting inside it the inner core and outer cladding.

9.1.2 Propagation of Optical Wave Inside the Fiber Optic Structure

Let us now consider the cylindrical dielectric structure as shown in Figure 9.3. This is just the geometry of the optical fiber, where the central region is known as the *core* and the outer region as the *cladding*. In this case, the basic principles are the same as for the dielectric slab, but the circular rather than planar symmetry changes the mathematics. We use the solution of Maxwell's equation in the cylindrical coordinates both for the coaxial cable and the circular waveguide, where we deal mostly with guiding modes rather than the ray concept [1–3, 6].

The wave equation that describes such propagation of light within cylindrical waveguides can be presented in cylindrical coordinates as follows for $\mu_r = 1$:

$$\nabla^2 E \equiv \frac{1}{r}\frac{\partial}{\partial r}\left(r\frac{\partial E}{\partial r}\right) + \frac{1}{r^2}\frac{\partial^2 E}{\partial \phi^2} + \frac{\partial^2 E}{\partial z^2} = \mu\varepsilon\frac{\partial^2 E}{\partial t^2} \tag{9.1}$$

We can present, as usually present in Refs. [1–3, 6], the solution taking into account separation of variables:

$$E = E_r(r)E_\phi(\phi)E_z(t) \tag{9.2}$$

From the well-known physics, we immediately take $E_t(t)E_z(t) = \exp\{i(\beta z - \omega t)\}$. This allows us to rewrite the wave equation Eq. (9.1) in the form

$$\frac{1}{r}\frac{\partial}{\partial r}\left(r\frac{\partial(E_rE_\phi)}{\partial r}\right) + \frac{1}{r^2}\frac{\partial^2(E_rE_\phi)}{\partial\varphi^2} - \beta^2 E_rE_\phi + \frac{n^2\omega^2}{c^2}(E_rE_\phi) = 0 \tag{9.3}$$

We now suggest that function E_ϕ is periodic and can be presented in the form

$$E_\phi = \exp(\pm im\phi) \tag{9.4}$$

where m is an integer. Now, we can reduce the above equation as

$$\frac{\partial^2 E_r}{\partial r^2} + \frac{1}{r}\frac{\partial E_r}{\partial r} + \left(n^2\frac{\omega^2}{c^2} - \beta^2 - \frac{m^2}{r^2}\right)E_r = 0 \tag{9.5}$$

Equation (9.5) is a form of Bessel's equation and its solutions are Bessel functions. We can finally obtain solutions for the field of rays through the modified Bessel functions of first and second order, $J(qr)$ and $K(pr)$, via wave parameters q and p as propagation parameters inside the core and cladding, respectively. This finally gives at the core $(r \le a)$

$$\frac{\partial^2 E_r}{\partial r^2} + \frac{1}{r}\frac{\partial E_r}{\partial r} + \left(q^2 - \frac{m^2}{r^2}\right)E_r = 0 \tag{9.6a}$$

and at the cladding $(r > a)$

$$\frac{\partial^2 E_r}{\partial r^2} + \frac{1}{r}\frac{\partial E_r}{\partial r} + \left(p^2 + \frac{m^2}{r^2}\right)E_r = 0 \tag{9.6b}$$

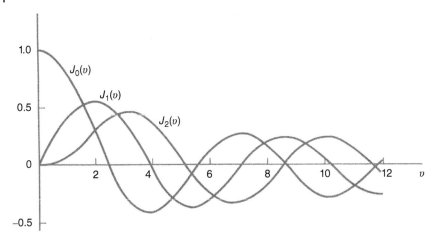

Figure 9.4 Bessel function of the first kind and *n* order vs. variable *v*.

Solutions of these equations, respectively, are

$$E_r = E_c J_m(qr) \tag{9.7a}$$

$$E_r = E_{c1} K_m(qr) \tag{9.7b}$$

where $J_m(qr)$ and $K_m(qr)$ are the Bessel function of the first kind and the modified Hankel function (Bessel function of the second kind), respectively. Roots of $J_m(qr) \equiv J_m(v)$, $m = 0, 1, 2, \ldots$, are shown in Figure 9.4.

Information on $K_m(qr)$ can be found, for example, in Refs. [8, 9]. Finally, the full solution at the core is

$$E = E_c J_m(qr) \exp\{-j(\omega t - \beta z)\} \exp(\pm jm\phi) \tag{9.8a}$$

and a similar full solution for the cladding is

$$E = E_{c1} K_m(qr) \exp\{-j(\omega t - \beta z)\} \exp(\pm jm\phi) \tag{9.8b}$$

where *l* is the azimuth integer.

As in the case of the 3-D classical empty waveguide, described in [1–5], we can determine for a fiber the corresponding values for the given propagation parameters *p*, *q*, and *β*, by imposing the boundary conditions at $r = a$. The result is a relationship that provides the *β* vs. *k* or *dispersion curves* shown in Figure 9.5.

The full mathematical approach is very complicated, and we use the so-called "weakly guiding" approximation. This makes use of the fact that if $n_1 \approx n_2$ the ray's angle of incidence at the boundary of core–cladding must be very large, if TIR is to occur. The ray must bounce down the core almost at grazing incidence. This means that the wave is very nearly a transverse wave, with very small *z*-components.

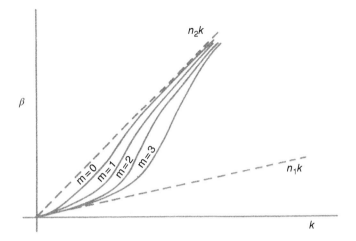

Figure 9.5 Dispersion diagram of optical modes in fiber structure.

Since the wave within fiber is considered to be transverse, the solution can be resolved conveniently into two linearly polarized (LP) components, as for free-space propagation. The modes are thus called *LP* modes [1–3, 6]. All solutions obtained above relate directly to the optical fiber guiding structures. The latter has just the cylindrical geometry, and if for a typical fiber we have that $(n_1 - n_2)/n_1 \approx 0.01$ then the "weakly guiding" approximation is valid.

There are two possible LP optical fiber modes: LP_{01} ($m = 0$) and LP_{11} ($m = 1$) [1–3, 6]. For cylindrical geometry the *single-mode condition* is [1–3]

$$\frac{2\pi a}{\lambda}(n_1^2 - n_2^2)^{1/2} < 2.404 \tag{9.9}$$

As follows from the illustrations presented, depending on the shape of intrinsic refractive index distribution, the corresponding LP modes can propagate asymmetrically and inhomogeneously. This phenomenon is called the *modal dispersion* [1–6] and will be discussed in Chapter 10.

References

1 Adams, M.J. (1981). *An Introduction to Optical Waveguides*. New York: Wiley.
2 Elliott, R.S. (1993). *An Introduction to Guided Waves and Microwave Circuits*. New Jersey: Prentice Hall.
3 Palais, J.C. (1998). *Fiber Optic Communications*, 4e. New Jersey: Prentice Hall.
4 Jackson, J.D. (1962). *Classical Electrodynamics*. New York: Wiley.

5 Chew, W.C. (1995). *Waves and Fields in Inhomogeneous Media*. New York: IEEE Press.

6 Dakin, J. and Culshaw, B. (eds.) (1988). *Optical Fiber Sensors: Principles and Components*. Boston-London: Artech House.

7 Palais, J.C. (2006). Optical communications. In: *Engineering Electromagnetics Applications* (ed. R. Bansal), 665–671. New York: Taylor & Frances.

8 Korn, G. and Korn, T. (1961). *Mathematical Handbook for Scientists and Engineers*. New York: McGraw-Hill.

9 Abramowitz, M. and Stegun, I.A. (1965). *Handbook of Mathematical Functions*. New York: Dover Publications.

10

Dispersion Properties of Fiber Optic Structures

To understand dispersive properties of fiber optic structures that transmit optical continuous or pulse signals at long distances, propagating along such optical waveguides, different kinds of which have been discussed in Chapter 9, let us introduce, according to Refs. [1–7], several important engineering parameters that characterize the corresponding processes of optical communication via fiber links.

10.1 Characteristic Parameters of Fiber Optic Structures

In fiber optics, there is an important parameter usually used, called *numerical aperture* of fiber optic guiding structure, denoted as NA [6, 7]

$$NA = n_1 \sin \theta_c \equiv \sin \theta_a \qquad (10.1)$$

where $\theta_{full} = 2 \cdot \theta_a$ is defined in the literature as the angle of minimum light energy spread outside the cladding or *of* full communication [1–7], when total internal reflection (TIR) occurs in the fiber optic structure. Accounting for $\cos^2\theta = 1 - \sin^2\theta$, we finally get

$$NA = (n_1^2 - n_2^2)^{1/2} \qquad (10.2)$$

Sometimes, in fiber optic physics, designers use the parameter, *relative refractive index* difference [1–7, 8]:

$$\Delta = \frac{(n_1^2 - n_2^2)}{2 \cdot n_1^2} \equiv \frac{(NA)^2}{2 \cdot n_1^2} \qquad (10.3)$$

Using the above formulas, we can find relations between these two engineering parameters:

$$NA = n_1 \cdot (2\Delta)^{1/2} \qquad (10.4)$$

Fiber Optic and Atmospheric Optical Communication, First Edition.
Nathan Blaunstein, Shlomo Engelberg, Evgenii Krouk, and Mikhail Sergeev.
© 2020 John Wiley & Sons, Inc. Published 2020 by John Wiley & Sons, Inc.

10.2 Dispersion of Optical Signal in Fiber Optic Structures

A problem of transmission of pulses via fiber optic structure occurs because of two factors. One is that the source of light is not emitted at a single wavelength but exists over a range of wavelengths called the source spectral width [1–7]. The second factor is that the index of refraction is not the same at all wavelengths. This property when the light velocity is dependent on wavelength is called *dispersion*.

10.2.1 Material Dispersion

We can call material dispersion as the property that accounts for the properties of the material from which fiber structures are developed. Such a kind of dispersion causes the spread of light wavelength as it travels along the fiber.

This is because each of the component wavelengths (called also *wave harmonic*) travels at a different speed, arriving with a slight delay with respect to the others. The amount of pulse spreading (τ) per unit length of fiber (l) is given by [6, 7]:

$$\Delta \left(\frac{\tau}{l} \right) = -M \Delta \lambda \tag{10.5}$$

where M is the *material dispersion factor*, plotted in Figure 10.1, according to [6, 7], for pure silica glass versus wavelength varied from 0.7 μm in units of picoseconds per nanometer per kilometer (ps/(nm · km)) of length of fiber.

It is clearly seen that $M = 0$ near 1300 nm, that is, here the pulse has a minimum spreading factor. In the range 1200–1600 nm, the material dispersion factor can be approximated by

$$M = \frac{M_0}{4} \left(\lambda - \frac{\lambda_0^4}{\lambda^3} \right) \tag{10.6}$$

where $M_0 \approx -0.095$ ps/(nm · km) and $\lambda_0 \approx 1300$ nm (wavelength where $M = 0$). At 1500 nm, $M \approx -20$ ps/(nm · km). For example, using an optical detector

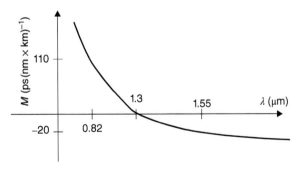

Figure 10.1 Dependence of dispersion properties of fiber optic materials vs. the length of optical wave λ (in μm), (it was taken from [6, 7]).

of the light-emitting diode (LED) type (see Chapter 8) with spectral width of 20 nm yields a pulse spread per unit length of the transmission path inside the fiber:

$$\Delta\left(\frac{\tau}{l}\right) = -M\Delta\lambda = -(-20) \cdot 20 = 400 \text{ ps/km} \tag{10.7}$$

10.2.2 Modal Dispersion

Now, we will discuss about the modal dispersion caused by nonsymmetrical distribution of the refractive index within the asymmetric fiber structure. In the case the *step-index fiber*, described earlier (see Chapter 9), following [6, 7], we can obtain a pulse spread per unit length along the fiber:

$$\Delta\left(\frac{\tau}{l}\right) = \frac{n_1}{cn_2}(n_1 - n_2) \tag{10.8}$$

If we account for the fractional refractive index, we finally get that the modal pulse spread as

$$\Delta\left(\frac{\tau}{l}\right) = \frac{n_1\Delta}{c} \tag{10.9}$$

Here, as above, c is the light speed. The problem with pulse spreading is that it limits the information-carrying capacity of the fiber. This aspect will be discussed in Chapter 11, where the data stream parameters of information passing the atmospheric and fiber optic channels will be considered.

The problem with pulse spreading is that it limits the information-carrying capacity of the fiber [7]. Pulses that spread eventually overlap with neighboring pulses, creating inter-symbol interference [7]. This leads to transmission errors and must be avoided. One direct way to avoid this is to place pulses further apart in the transmitter. This means lowering the data rate. The limits on data capacity caused by pulse spreading for non-return-to-zero (NRZ) pulse codes (see definitions in Chapters 1 and 5) is

$$C_{\text{NRZ}} \times l = \frac{0.7}{\Delta(\tau/l)} \tag{10.10a}$$

and for return-to-zero (RZ) pulse codes (see definitions in Chapters 1 and 5) is

$$C_{\text{RZ}} \times l = \frac{0.35}{\Delta(\tau/l)} \tag{10.10b}$$

Using the numerical values in the preceding example described by (10.10), we get, respectively,

$$C_{\text{NRZ}} \times l = 1.75 \text{ Mbps} \times \text{km} \tag{10.11a}$$

and

$$C_{\text{RZ}} \times l = 0.875 \text{ Mbps} \times \text{km} \tag{10.11b}$$

Similarly, pulse spreading reduces the bandwidth of an analog system. In such a system, the 3-dB bandwidth limit is

$$f|_{3\,dB} \times l = \frac{0.35}{\Delta\tau/\Delta L} \qquad (10.12a)$$

or using the same numerical values from the example described by (10.10b) and (10.11b), we get the limit in bandwidth

$$f|_{3\,dB} \times l = 0.875\,\text{MHz} \times \text{km} \qquad (10.12b)$$

At modulation frequencies much lower than that calculated above, the analog signal propagates without distortion. At the 3-dB frequency the amplitude of the signal diminishes to 50% of what it was at lower frequencies. At modulation frequencies well above the 3-dB value, the signals are attenuated greatly. Pulse spreading causes the fiber to act as a low-pass filter, allowing only the lower modulating frequencies to pass. Another kind of dispersion is the *multimode dispersion* occurring in the *step-index fiber* (see Chapter 9), when different rays travel along paths of different lengths. In this case,

$$\Delta T = \frac{l \cdot n_1^2}{c n_2}\Delta \sim \frac{l \cdot n_1}{c}\Delta \qquad (10.13)$$

The same *multimode dispersion* occurs also when oblique rays pass the *graded-index* (GRIN) fiber (see Chapter 9), which has longer path, but lower refractive index. The field distribution along the core of such a fiber with the Gaussian variant of parabolic GRIN is shown in Figure 10.2 according to [6, 7].

The fiber with such a kind of refractive index distribution has the minimum multimode dispersion. Here, the modal distortion can be found by the following

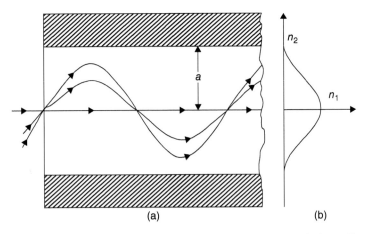

(a)   (b)

Figure 10.2 Field distribution inside GRIN fiber with graded parabolic profile of the refractive index inside the core (a) (rearranged from [6, 7]).

expression [6, 7]:

$$\Delta\left(\frac{\tau}{l}\right) = \frac{n_1\Delta^2}{2c} \tag{10.14}$$

It can be seen that GRIN fiber gives reduction in pulse spread (and resultant increase in fiber capacity) compared with step-index fiber close to a factor of 200. Therefore, GRIN fiber is usually used when path lengths are moderate (such as in LAN applications), where path length up to a few kilometers and information rates of a few gigabits per second (Gbps) can be accommodated. Longer paths and higher rates require single-mode fibers, where modal distortion is no longer a factor.

We should mention that with deviation of polarization of optical waves inside the fiber optic, *polarization mode dispersion* (PMD) may have occurred. It is characterized by the following characteristics:

$$\sigma_p = D_p\sqrt{l} \tag{10.15}$$

Here, D_p is the PMD parameter, measured in picoseconds per square root of kilometer, which for current fiber is less than $0.5\,\text{ps/km}^{1/2}$ and for many installed spans can exceed $10\,\text{ps/km}^{1/2}$.

Finally, we will mention attenuation of light wave energy inside the fiber. Losses inside the fiber structure, as was described in Chapter 9, are determined by factor α, called *attenuation coefficient*, determined by *Nepers per meter* (*Np/m*). Typically, fiber attenuation is given in *dB/km*. If so, we can present relations between these two parameters as

$$dB/km = -8.685 \cdot \alpha \tag{10.16}$$

where the units of the attenuation coefficient are km^{-1}. As was shown in Ref. [7], the wavelength regions where the fiber optic system has the absorption peak are at the wavelength of around 1300 nm and a side peak at the wavelength range of 940–980 nm, and have transparent "windows" for optic signal transmission with data, which are at the regions around 800–900 nm and from 1250 to 1650 nm.

Problems

10.1 Given a fiber optical structure with $n_1 = 1.45$ and $\Delta = 0.02(2\%)$. Find: NA and $2 \cdot \theta_a$.

10.2 Given the refractive index of core (with radius a) $n_1 = 1.48$, and of cladding (with width $b-a$) $n_2 = 1.465$ (see Figure 10.3). The outer layer (with width $c-b$) has the same refractive index as the core. Find: Fractional refractive index Δ and modal distortion.

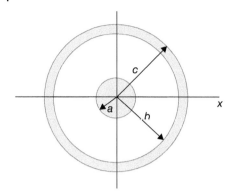

Figure 10.3 View of Problem #2.

10.3 Given optical fiber with the refractive index of core $n_1 = 1.48$, and of cladding $n_2 = 1.465$ (see Figure 10.3).
Find: Fractional refractive index and modal distortion.

References

1 Adams, M.J. (1981). *An Introduction to Optical Waveguides*. New York: Wiley.
2 Elliott, R.S. (1993). *An Introduction ro Guided Waves and Microwave Circuits*. New Jersey: Prentice Hall.
3 Palais, J.C. (1998). *Fiber Optic Communications*, 4e. New Jersey: Prentice-Hall.
4 Jackson, J.D. (1962). *Classical Electrodynamics*. New York: Wiley.
5 Chew, W.C. (1995). *Waves and Fields in Inhomogeneous Media*. New York: IEEE Press.
6 Dakin, J. and Culshaw, B. (eds.) (1988). *Optical Fiber Sensors: Principles and Components*. Boston-London: Artech House.
7 Palais, J.C. (2006). Optical communications. In: *Handbook of Engineering Electromagnetics Applications* (ed. R. Bansal). New York: Taylor & Frances.

Part IV

Wireless Optical Channels

11

Atmospheric Communication Channels

11.1 Basic Characteristics of Atmospheric Channel

The atmosphere is a gaseous envelope that surrounds the Earth from the ground surface up to several hundred kilometers. The atmosphere consists of different kinds of gaseous, liquid, and crystal structures, including effects of gas molecules (atoms), aerosol, cloud, fog, rain, hail, dew, rime, glaze, and snow [1–16]. Except for the first two, the others are usually called *hydrometeors* in the literature [9–15]. Furthermore, due to irregular and sporadic air streams and motions, that is, irregular wind motions, chaotic structures defined as *atmospheric turbulence* are also present in the atmosphere [17–21].

Based mostly on temperature variations, the Earth's atmosphere is divided into four primary layers [20]:

(1) the *troposphere* that surrounds the Earth from the ground surface up to 10–12 km above the terrain, with the *tropopause* region as the isothermal layer above the troposphere up to ~20 km; it continuously spreading to the stratosphere;
(2) the *stratosphere* with the stratopause (from 20 km up to ~50 km altitude);
(3) the *mesosphere* with the mesopause (up to ~90 km);
(4) the *thermosphere* (up to ~600 km), containing the multilayered plasma structure, usually called ionosphere (70–400 km).

In our further discussions, we will briefly focus on the effects of the *troposphere* on optical wave propagation starting with a definition of the troposphere as a natural layered air medium consisting of different gaseous, liquid, and crystal structures.

The physical properties of the atmosphere are characterized by main parameters such as *temperature*, T (in kelvin), *pressure*, p (in millibars, *pascals* or in mm Hg), and *density*, ρ (in kg/m^3). All these parameters change significantly with altitude and seasonal and latitudinal variability, and strongly depend on environmental conditions occurring in the troposphere [22].

Fiber Optic and Atmospheric Optical Communication, First Edition.
Nathan Blaunstein, Shlomo Engelberg, Evgenii Krouk, and Mikhail Sergeev.
© 2020 John Wiley & Sons, Inc. Published 2020 by John Wiley & Sons, Inc.

Over 98% of the troposphere is comprised of the elements nitrogen and oxygen. The number density of nitrogen molecules, $\rho_N(h)$, vs. the height h can be found in Ref. [22]. The temperature $T(h)$ and the pressure $P(h)$, as a function of the altitude h (in meters), for the first 11 km of the troposphere can be determined from the expressions (11.1) and (11.2) [6, 23]:

$$T(h) = 288.15 - 65.45 \cdot 10^{-4} h \tag{11.1}$$

$$P(h) = 1.013 \times 10^5 \cdot \left[-\frac{288.15}{T(h)} \right]^{5.22} \tag{11.2}$$

The temperature and pressure from 11 to 20 km in the atmosphere can be determined from

$$T(h) = 216.65 \tag{11.3}$$

$$P(h) = 2.269 \times 10^4 \cdot \exp\left[-\frac{0.034\,164(h - 11\,000)}{216.65} \right] \tag{11.4}$$

The number density of molecules can be found from [6]:

$$\rho(h) = \left(\frac{28.964 \text{ kg/kmol}}{8314 \text{ J/kmol} - K} \right) \cdot \frac{P(h)}{T(h)} = 0.003\,484 \cdot \frac{P(h)}{T(h)} \text{ kg/m}^3 \tag{11.5}$$

11.2 Effects of Aerosols on Atmospheric Communication Links

Aerosol is a system of liquid or solid particles uniformly distributed in the atmosphere [13, 24–41]. Aerosol particles play an important role in the precipitation process, providing the nuclei upon which condensation and freezing take place. The particles participate in chemical processes and influence the electrical properties of the atmosphere.

11.2.1 Aerosol Dimensions

Actual aerosol particles range in diameter from a few nanometers to about a few micrometers. When smaller particles are in suspension, the system begins to acquire the properties of a real aerosol structure. For larger particles, the settling rate is usually so rapid that the system cannot properly be called a real aerosol. Nevertheless, the term is commonly employed, especially in the case of fog or cloud droplets and dust particles, which can have diameters of over 100 μm. In general, aerosols composed of particles larger than about 50 μm are unstable unless the air turbulence is extreme, as in a severe thunderstorm (see details in Refs. [5–7]).

Of all the classifications of atmospheric aerosols the most commonly used one is according to size. The general classification suggests three modes of aerosols [13]:

(1) a *nuclei* mode that is generated by spontaneous nucleation of the gaseous material for particles less than 0.1 μm in diameter;
(2) the *accumulation* mode for particles between 0.1 and 1 μm diameter, mainly resulting from coagulation and in cloud processes; and
(3) the *coarse* mode for particles larger than 1.0 μm in diameter originating from the Earth's surface (land and ocean).

The classification is quite similar to the Junge's designation [24] referring to aerosols as Aitken, large, and giant particles. The particles vary not only in chemical composition and size but also in shape (spheres, ellipsoids, rods, etc.).

Aerosol concentrations and properties depend on the intensity of the sources, on the atmospheric processes that affect them, and on the particle transport from one region to another. The size distribution of the atmospheric aerosol is one of its core physical parameters. It determines the various properties such as mass and number density, or optical scattering, as a function of particle radius. For the atmospheric aerosols this size range covers more than 5 orders of magnitude, from about 10 nm to several hundred micrometers. This particle size range is very effective for scattering of radiation at ultraviolet (UV), visible, and infrared (IR) wavelengths. The aerosol size distribution varies from place to place, with altitude, and with time. In a first attempt to sort into geographically distinct atmospheric aerosols, Junge classified aerosols depending on their location in space and sources into background, maritime, remote continental, and rural [24]. This classification was later expanded and quantified [25–30].

11.2.2 Aerosol Altitudes Localization

Because the aerosol in the atmosphere exhibits considerable variation in location, height, time, and constitution, different concepts exist for describing the aerosol loading in the atmosphere. Models for the vertical variability of atmospheric aerosols are generally broken into a number of distinct layers. In each of these layers a dominant physical mechanism determines the type, number density, and size distribution of particles. The generally accepted layer models consist of the following [33, 34]:

(1) a boundary layer that includes aerosol mixing goes from 0 to 2–2.5 km elevation;
(2) free tropospheric region running from 2.5 to 7–8 km;
(3) a stratospheric layer from 8 to 30 km; and
(4) layers above 30 km composed mainly of particles that are extraterrestrial in origin such as meteoric dust [21].

The average thickness of the aerosol-mixing region is approximately 2–2.5 km. Within this region, one would expect the aerosol concentration to be influenced strongly by conditions at ground level. Consequently, aerosols in this region display the highest variability with meteorological conditions, climate, etc. [34–41].

11.2.3 Aerosol Concentration

In the free tropospheric layer that extends from 2–2.5 to 7–8 km, an exponential decay of aerosol number density is observed. The total number of their density varies as [12]

$$N(z) = N(0) \exp\left(-\frac{z}{z_s}\right) \tag{11.6}$$

where the scale height z_s ranges from 1 to 1.4 km.

The little data available show that in the atmospheric boundary layer (0–2 km) and in the lower stratosphere (9–14 km altitude) different layers (known as mixing layers) can exist with constant and increased aerosol concentration [36, 37]. These layers can be caused by temperature inversions at ground level and by tropopause effects where the temperature gradient changes sign. The concentration maximum of stratospheric aerosols near the equator is located at 22–26 km altitude, but at about 17–18 km height in the polar region.

11.2.4 Aerosol Size Distribution and Spectral Extinction

The number $n(r)$ of particles per unit interval of radius and per unit volume is given by

$$n(r) = \frac{dN(r)}{dr} \tag{11.7}$$

The differential quantity $dN(r)$ expresses the number of particles having a radius between r and $r + dr$, per unit volume, according to the distribution function $n(r)$.

Because of the many orders of magnitude present in atmospheric aerosol concentrations and radii, a logarithmic size distribution function is often used:

$$n(r) = \frac{dN(r)}{d\log(r)} \tag{11.8}$$

The much used distribution function is the power law first presented by Junge [24, 40]. The Junge's model is

$$n(r) = \frac{dN(r)}{d\log r} = C \cdot r^{-\nu}, \quad r \geq r_{min} \tag{11.9}$$

or in a non-logarithmic form

$$n(r) = \frac{dN(r)}{dr} = 0.434 C r^{-v} \tag{11.10}$$

where C is the normalizing constant to adjust the total number of particles per unit volume and v is the shaping parameter. Most measured size distributions can best be fit by values of v in the range $3 \leq v \leq 5$, for hazy and clear atmospheric conditions, and for aerosols whose radii lie between 0.1 and 10 μm [41]. According to the power-law size distribution, the number of particles decreases monotonically with an increase in radius. In practice, there is an accumulation in the small particle range. Actual particle size distributions may differ considerably from a strict power-law form.

The modified power-law distribution was given by McClatchey *et al.* [41], which was then modified in Ref. [42] by the use of Gamma–Gamma PDF (see definitions in Chapter 6):

$$n(r) = \frac{dN(r)}{dr} = a r^{\alpha} \exp(-b r^{\beta}) \tag{11.11}$$

where a is the total number density and α, β, and b are the shaping parameters.

The total particle concentration, given by the integral over all particle radii, for this distribution according to Ref. [39] is

$$N = a \cdot \beta^{-1} \cdot b^{-(\alpha+1)/\beta} \Gamma\left(\frac{\alpha+1}{\beta}\right) \tag{11.12}$$

The mode radius for this distribution is given by [6, 39]

$$r_m^{\alpha} = \frac{\alpha}{b \cdot \beta} \tag{11.13}$$

The value of the distribution at the mode radius is [6, 39]

$$n(r_m) = a \cdot r_m^{\alpha} \cdot \exp(-\alpha/\beta) \tag{11.14}$$

Because it has four adjustable constants, (11.14) can be fitted to various aerosol models. The Gamma PDF is usually employed to model haze, fog, and cloud particle size distributions.

The tropospheric aerosols above the boundary layer are assumed to have the same composition, but their size distribution is modified by eliminating the large particle component.

It is often found that within the optical subrange (~0.16 to 1.2 μm) the light scattering coefficient, $\alpha(\lambda)$, and aerosol size distribution in form (11.13) obey the following power-law relationship [12]:

$$\alpha(\lambda) = C \cdot \lambda^{-b} \tag{11.15}$$

where b is referred to as the exponent power that equals $b = v - 3$ [12]. Thus, if α_{λ} depends strongly on wavelength (large b), then the size distribution function

(11.15) decreases with particle size. Measurements have shown that values of b tend to be higher for continental aerosols than for clean marine aerosols [6, 12].

Multiple aerosol scattering induces loss of optical wave after it has traversed the scatters. In optical communication, the dense aerosol/dust layers, as part of the wireless atmospheric communication channel, can cause signal power attenuation, as well as temporal and spatial signal fluctuations (i.e. fading; see definitions in Chapter 6). This effect limits the maximum data rate and increases the bit error rate (BER). Thus, knowledge of the parameters that determine the optical properties of atmospheric aerosols (spectral extinction and size distribution) is essential for development of techniques for wireless optical communications through the atmosphere (see Chapter 12).

11.3   Effects of Hydrometeors

Hydrometeors are any water or ice particles that have formed in the atmosphere or at the Earth's surface as a result of condensation or sublimation. Water or ice particles blown from the ground into the atmosphere are also classed as hydrometeors. Some well-known hydrometeors are rain, fog, snow, clouds, hail and dew, glaze, blowing snow, and blowing spray. Scattering by hydrometeors has an important effect on signal propagation.

11.3.1   Effects of Fog

Fog is a cloud of small water droplets near ground level and is sufficiently dense to reduce horizontal visibility to less than 1000 m. Fog is formed by the condensation of water vapor on condensation nuclei that are always present in natural air. This can occur as soon as the relative humidity of the air exceeds saturation by a fraction of 1%. In highly polluted air the nuclei may grow sufficiently to cause fog at humidities of 95% or less. Three processes can increase the relative humidity of the air:

(1) cooling of the air by adiabatic expansion;
(2) mixing two humid airstreams having different temperatures; and
(3) direct cooling of the air by radiation (namely, cosmic ray radiation).

According to the physical processes of fog creation, there are different kinds of fogs usually observed: advection, radiation, inversion, and frontal. We do not enter deeply into the subject of their creation, referring the reader to special literature [5–7].

When the air becomes nearly saturated with water vapor (RH $\rightarrow$ 100 %), fog can form assuming that sufficient condensation nuclei are present. The air can become saturated in two ways, either by mixing of air masses with different temperatures and/or humidities (advection fogs), or by the air cooling until the air temperature approaches the dew point temperature (radiation fogs).

The fog models, which describe the range of different types of fog, have been widely presented based on measured size distributions [33]. The modified gamma size distribution (1.17) was used to fit the data. The models represent heavy and moderate fog conditions. The developing fog can be characterized by droplet concentrations of 100–200 particles per cubic centimeter in the 1–10 μm radius range with a mean radius of 2–4 μm. As the fog thickens, the droplet concentration decreases to less than 10 particles per cubic centimeter and the mean radius increases from 6 to 12 μm. Droplets less than 3 μm in radius are observed in fully developed fog. It is usually assumed that the refractive index of the fog corresponds to that of pure water. Natural fogs and low level clouds are composed of spherical water droplets, the refractive properties of which have been fairly well documented in the spectral region of interest.

11.3.2 Effects of Rain

Rain is the precipitation of liquid water drops with diameters greater than 0.5 mm. When the drops are smaller, the precipitation is usually called drizzle. Concentration of raindrops typically spreads from 100 to 1000 m^{-3}. Raindrops seldom have diameters larger than 4 mm because, as they increase in size, they break up. The concentration generally decreases as diameters increase. Meteorologists classify rain according to its rate of fall. The hourly rates relate to light, moderate, and heavy rain, which correspond to dimensions less than 2.5 mm, between 2.8 and 7.6 mm, and more than 7.6 mm, respectively. Less than 250 mm and more than 1500 mm per year represent approximate extremes of rainfall for all of the continents (see details in [1, 6–10]).

Thus, if such parameters of rain as the density and size of the drops are constant, then, according to [6–10], the signal power P_r at the receiver decreases exponentially with the optical ray path r, through the rain, with the parameter of power attenuation in e^{-1} times, α; that is,

$$P_r = P_r(0) \exp\{-\alpha r\} \tag{11.16}$$

Expressing (11.16) in logarithmic scale gives

$$L = 10 \log \frac{P_t}{P_r} = 4.343 \alpha r \tag{11.17}$$

Another way to estimate the total loss via the specific attenuation in decibels per meter was shown by Saunders [10]. He defined this factor as

$$\gamma = \frac{L}{r} = 4.343 \alpha \tag{11.18}$$

where now the power attenuation factor α can be expressed through the integral effects of the one-dimensional (1-D) diameter D of the drops, defined by $N(D)$, and the effective cross-section of the frequency-dependent signal power attenuation by rain drops, $C(D)$ (dB/m), as

$$\alpha = \int_{D=0}^{\infty} N(D) \cdot C(D) dD \tag{11.19}$$

As was mentioned in Refs. [8–10], in real tropospheric situations, the drop size distribution $N(D)$ is not constant and can be accounted for by the range dependence of the specific attenuation – that is, range dependence, $\gamma = \gamma(r)$ – and integrating it over the whole optical ray path length r_R gives the total path loss

$$L = \int_{0}^{r_R} \gamma(r) dr \tag{11.20}$$

To resolve Eq. (11.19), a special mathematical procedure was proposed in [10] that accounts for the drop size distribution. This procedure yields an expression for $N(D)$ as

$$N(D) = N_0 \exp\left\{-\frac{D}{D_m}\right\} \tag{11.21}$$

where $N_0 = 8 \cdot 10^3$ m^{-2} mm^{-1} is a constant parameter [10], and D_m is the parameter that depends on the rainfall rate R, measured above the ground surface in millimeters per hour, as

$$D_m = 0.122 \cdot R^{0.21} \text{ mm} \tag{11.22}$$

As for the attenuation cross section $C(D)$ from (11.19), it can be found using the so-called Rayleigh approximation that is valid for lower frequencies, when the average drop size is small compared to the radio wavelength. In this case, only absorption inside the drop occurs and the Rayleigh approximation is valid, giving a very simple expression for $C(D)$, i.e.

$$C(D) \propto \frac{D^3}{\lambda} \tag{11.23}$$

Attenuation caused by rain increases more slowly with frequencies, approaching a constant value known as the *optical limit*. Near this limit, scattering forms a significant part of attenuation, which can be described using the *Mie* scattering theory discussed in Ref. [10].

In practical situations, an empirical model is usually used, where $\gamma(r)$ is assumed to depend only on rainfall rate R and wave frequency. Then according to [1, 9, 10], we can obtain

$$\gamma(f, R) = a(f) R^{b(f)} \tag{11.24}$$

where γ has units dB/km; $a(f)$ and $b(f)$ depend on frequency (GHz).

The cell diameter appears to have an exponential probability distribution of the form [8–10]

$$P(D) = \exp(-D/D_0) \tag{11.25}$$

where D_0 is the mean diameter of the cell and is a function of the peak rainfall rate R_{peak}. For Europe and in the United States, the mean diameter D_0 decreases slightly with increasing R_{peak} when $R_{peak} > 100$ mm/h. This relationship appears to obey a power law

$$D_0 = aR_{peak}^{-b}, \quad R_{peak} > 10\,\text{mm/h} \tag{11.26}$$

Values for the coefficient a ranging from 2 to 4, and the coefficient b from 0.08 to 0.25 have been reported.

11.3.3 Effects of Clouds

Clouds have dimensions, shape, structure, and texture, which are influenced by the kind of air movements that result in their formation and growth, and by the properties of the cloud particles. In settled weather, clouds are well scattered and small and their horizontal and vertical dimensions are only a kilometer or two. In disturbed weather they cover a large part of the sky, and individual clouds may tower as high as 10 km or more. Growing clouds are sustained by upward air currents, which may vary in strength from a few cm/s to several m/s. Considerable growth of the cloud droplets, with falling speeds of only about one cm/s, leads to their fall through the cloud and reaching the ground as drizzle or rain. Four principal classes are recognized when clouds are classified according to the kind of air motions that produce them:

(1) layer clouds formed by the widespread regular ascent of air;
(2) layer clouds formed by widespread irregular stirring or turbulence;
(3) cumuliform clouds formed by penetrative convection;
(4) orographic clouds formed by ascent of air over hills and mountains.

We do not enter into the subject of how such kinds of clouds are created, because it is a matter of meteorology. Interested readers can find information in [1–8].

There have been several proposed alternative mathematical formulations for the probability distribution of sky cover, as an observer's view of the sky dome. Each of them uses the variable x ranging from 0 (for clear conditions) to 1.0 (for overcast). We do not enter into a description of these models because they are fully described in Refs. [4, 6–8].

In cloud models, described in [4, 6–8], a distinction between *cloud cover* and *sky cover* should be explained. Sky cover is an observer's view of cover of the sky dome, whereas cloud cover can be used to describe areas that are smaller or larger than the floor space of the sky dome. It follows from numerous observations that in clouds and fog the drops are always smaller than 0.1 mm, and the theory for the small size scatterers is applicable [7, 20–24]. This gives the attenuation coefficient

$$\gamma_c \approx 0.438c(t)q/\lambda^2, \quad \text{dB/km} \tag{11.27}$$

where λ is the wavelength measured in centimeters and q is the water content measured in grams per cubic meter. For the visibility of 600, 120, and 30 m the water content in fog or cloud is 0.032, 0.32, and 2.3 g/m^3, respectively. The calculations show that the attenuation in a moderately strong fog or cloud does not exceed the attenuation due to rain with a rainfall rate of 6 mm/h.

11.3.3.1 Snow

Snow is the solid form of water that crystallizes in the atmosphere and, falling to the Earth, covers permanently or temporarily about 23% of the Earth's surface. Snow falls at sea level poleward of latitude 35° N and 35° S, although on the west coast of continents it generally falls only at higher latitudes. Close to the equator, snowfall occurs exclusively in mountain regions, at elevations of 4900 m or higher. The size and shape of the crystals depend mainly on the temperature and the amount of water vapor available as they develop. In colder and drier air, the particles remain smaller and compact. Frozen precipitation has been classified into seven forms of snow crystals and three types of particles: graupel, i.e. granular snow pellets, also called soft hail; sleet, that is, partly frozen ice pellets; and hail, e.g. hard spheres of ice (see details in [1–4, 6–9]).

11.4 Effects of Turbulent Gaseous Structures on Optical Waves Propagation

The temperature and humidity fluctuations combined with turbulent mixing by wind and convection induce random changes in the air density and in the field of atmospheric refractive index in the form of optical turbules (or eddies), called optical turbulence, which is one of the most significant parameters for optical wave propagation [35–47]. Random space–time redistribution of the refractive index causes a variety of effects on an optical wave related to its temporal irradiance fluctuations (scintillations) and phase fluctuations. A statistical approach is usually used to describe both atmospheric turbulence and its various effects on optical or IR systems.

11.4.1 Turbulence Phenomenon

Atmospheric turbulence is a chaotic phenomenon created by random temperature, wind magnitude variation, and direction variation in the propagation medium. This chaotic behavior results in index-of-refraction fluctuations. The turbulence spectrum is divided into three regions by two scale sizes [35–47]:

– the outer scale (or macro size) of turbulence: L_0;
– the inner scale (or micro size) of turbulence: l_0.

These values vary according to atmosphere conditions, distance from the ground, and other factors. The inner scale l_0 is assumed to lie in the range of 1–30 mm. Near the ground, it is typically observed to be around 3–10 mm, but generally increases to several centimeters with increasing altitude h. A vertical profile for the inner scale is not known. The outer scale L_0, near the ground, is usually taken to be roughly kh, where k is a constant on the order of unity and h is the atmospheric layer of turbulence localization. Thus, L_0 is usually either equal to the height from the ground (when the turbulent cell is close to the ground) or in the range of 10–100 m or more. Vertical profile models for the outer scale have been developed based on measurements, but different models predict very different results.

The main goal of studying optical wave propagation through a turbulent atmosphere is the identification of tractable *PDF*s and *CDF*s or the corresponding spectra of the irradiance under all irradiance fluctuation conditions. Obtaining an accurate mathematical model for a *PDF* and a *CDF* of the randomly fading irradiance signal will enable the link planner to predict the reliability of a radio communication system operating in such an environment. In addition, it is beneficial if the free parameters of that *PDF* and *CDF* can be tied directly to atmospheric parameters.

To investigate physical properties of turbulent liquid, at the earliest study of turbulent flow, Reynolds used the theory "of similarity" to define a nondimensional quantity $Re = V \cdot l/v$, called the Reynolds number [35–47], where V and l are the characteristic velocity (in m/s) and size (in m) of the flow respectively, and v is the kinematic viscosity (in m^2/s). The transition from laminar to turbulent motion takes place at a critical Reynolds number, above which the motion is considered to be turbulent. The kinematic viscosity v of air is roughly 10^{-5} m^2/s [5–7]; then air motion is considered highly turbulent in the boundary layer and troposphere, where the Reynolds number Re~10^5 [43–47].

Richardson [43] first developed a theory of the turbulent energy redistribution in the atmosphere – the energy cascade theory. It was noticed that smaller scale motions originated as a result of the instability of larger ones. A cascade process, shown in Figure 11.1 attracted from [44], in which eddies of the largest size are broken into smaller and smaller ones, continues down to scales in which the dissipation mechanism turns the kinetic energy of motion into heat.

Let us denote by l the current size of turbulence eddies, by L_0 and l_0 – their outer and inner scales – and by $\kappa_0 = \frac{2\pi}{L_0}$, $\kappa = \frac{2\pi}{l}$, and $\kappa_m = \frac{2\pi}{l_0}$ – the spatial wave numbers of these kinds of eddies – respectively. In these notations, one can

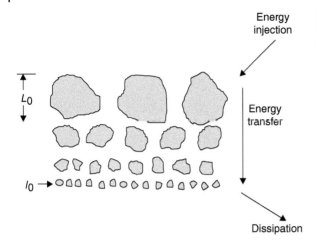

Energy
injection

Energy
transfer

Dissipation

Figure 11.1 Richardson's cascade theory of turbulence.

divide turbulences at the three regions:

$$\text{Input range:} \quad L_0 \leq l, \quad \kappa < \frac{2\pi}{L_0}$$

$$\text{Inertial range:} \quad l_0 < l < L_0, \quad \frac{2\pi}{L_0} < \kappa < \frac{2\pi}{l_0} \tag{11.28}$$

$$\text{Dissipation range:} \quad l \geq l_0, \quad \frac{2\pi}{l_0} \geq \kappa$$

These three regions induce strong, moderate, and weak spatial and temporal variations, respectively, of signal amplitude and phase, referred to in the literature as *scintillations* [1–12].

Kolmogorov [44, 45] introduced a hypothesis stating that during the cascade process the direct influence of larger eddies is lost and smaller eddies tend to have independent properties, universal for all types of turbulent flows. Following Kolmogorov, the energy cascade process consists of an energy *input region, inertial subrange,* and *energy dissipation region,* as sketched in Figure 11.1.

At large characteristic scale or eddy, a portion of kinetic energy in the atmosphere is converted into turbulent energy. When the characteristic scale reaches an outer scale size, L_0, the energy begins a cascade that forms a continuum of eddy size for the energy transfer from a macroscale L_0 to a microscale l_0 called the inner turbulence scale. The scale sizes l bounded above by L_0 and below by l_0 form the inertial subrange.

Kolmogorov proposed that in the inertial subrange, where $L_0 > l > l_0$, turbulent motions are both homogeneous and isotropic and energy may be transferred from eddy to eddy without loss, i.e. the amount of energy that is being injected into the largest structure must be equal to the energy that is dissipated as heat [1, 4].

When the size of a decaying eddy reaches l_0, the energy is dissipated as heat through viscosity processes. It was also hypothesized that the motion associated with the small-scale structure l_0 is uniquely determined by the kinematic viscosity v and ε, where $l_0 \sim \eta = (v^3/\varepsilon)^{1/4}$ is the Kolmogorov microscale, and ε is the average energy dissipation rate [44–47]. The Kolmogorov microscale defines the eddy size dissipating the kinetic energy. The turbulent process, shown schematically in Figure 11.1 according to the simple theory of Richardson, was then summarized by Kolmogorov and Obukhov (called the Kolmogorov–Obukhov turbulent cascade process in the literature [44–47]) as follows: the average energy dissipation rate ε of the turbulent kinetic energy will be distributed over the spatial wavelength κ-range as [44]

$$\text{In the input range} \quad (\kappa_0 \sim 1/L_0) : \varepsilon \sim \kappa_0^{-5/3}$$

$$\text{In the inertial range} \quad (\kappa \sim 1/l) : \varepsilon \sim \kappa^{-5/3} \tag{11.29}$$

$$\text{In the dissipation range} \quad (\kappa_m \sim 1/l_0) : \varepsilon \sim \kappa_m^{-5/3}$$

where, as above, L_0, l, and l_0 are the initial (outer), current, and inner turbulent eddy sizes.

In general, turbulent flow in the atmosphere is neither homogeneous nor isotropic. However, it can be considered locally homogeneous and isotropic in small subregions of the atmosphere. Finally, we should mention that atmospheric turbulences due to their motion can cause strong frequency-selective or flat fast fading (see definitions in Chapter 6). Below, we will analyze briefly some effects of the turbulent structures on optical signals/rays passing gaseous turbulent irregular atmosphere.

11.4.2 Scintillation Phenomenon of Optical Wave Passing the Turbulent Atmosphere

Optical waves, traveling through the turbulent atmosphere, which is characterized by rapid variations of refraction indexes (see previous sections), undergo fast changes in their amplitude and phase [8, 17–19, 35, 45]. This effect is called *dry tropospheric scintillation*. The phase and amplitude fluctuations occur both in the space and time domains. Moreover, this phenomenon is strongly frequency dependent: the shorter wavelengths lead to more severe fluctuations of signal amplitude and phase resulting from a given scale size [6, 20, 21]. The scale size can be determined by experimentally monitoring the scintillation of an optical signal on two nearby paths and/or by examination of the cross-correlation between the scintillations along the propagation paths. If the effects are closely correlated, then the scale size is large compared with the path spacing [48]. Additional investigations have shown that the distribution of the signal fluctuations (in decibels) is approximately a Gaussian distribution, whose standard deviation is the intensity [8, 17–21].

11.4.3 Scintillation Index

A wave propagating through a random medium such as the atmosphere will experience irradiance fluctuations, called scintillations, even over relatively short propagation paths. Scintillation is defined as [8, 17–19]

$$\sigma_1^2 = \frac{\langle I^2 \rangle - \langle I \rangle^2}{\langle I \rangle^2} = \frac{\langle I^2 \rangle}{\langle I \rangle^2} - 1 \tag{11.30}$$

This is caused almost exclusively by small temperature variations in the random medium, resulting in index-of-refraction fluctuations (i.e. turbulent structures). In (11.30) the quantity I denotes irradiance (or intensity) of optical wave and the angular brackets denote an ensemble average or equivalently, a long-time average. In weak fluctuation regimes, defined as those regimes for which the scintillation index is less than one [8, 17–19], derived expressions for the scintillation index show that it is proportional to *Rytov variance*:

$$\sigma_R^2 = 1.23 C_n^2 k^{7/6} R^{11/6} \tag{11.31}$$

Here, C_n^2 is the index-of-refraction structure parameter defined above, k is the radio wave number, and R is the propagation path length between transmitter and receiver.

The Rytov variance represents the scintillation index of an unbounded plane wave in the case of its weak fluctuations but is otherwise considered a measure of the turbulence strength when extended to strong fluctuation regimes by increasing either C_n^2 or the path length R, or both. It is shown in [8, 17–19] that the scintillation index increases with increasing values of the Rytov variance until it reaches a maximum value greater than unity in the regime characterized by random focusing, because the focusing caused by large-scale inhomogeneities achieves its strongest effect. With increasing path length or inhomogeneity strength, multiple scattering weakens the focusing effect, and the fluctuations slowly begin to decrease, saturating at a level for which the scintillation index approaches the value of one from above. Qualitatively, saturation occurs because multiple scattering causes the optical wave to become increasingly less coherent in the process of wave propagation through random media.

11.4.4 Signal Intensity Scintillations in the Turbulent Atmosphere

Early investigations concerning the propagation of unbounded plane waves and spherical waves through random media obtained results limited by weak fluctuations [48]. To explain the weak fluctuation theory, three new parameters of the problem must be introduced instead of the inner and outer scales of turbulences described earlier. They are

(a) the coherence scale, $l_1 \equiv l_{co} \sim 1/\rho_0$, which describes the effect of coherence between two neighboring points (see [8]);
(b) the first Fresnel zone scale, $l_2 \equiv \ell_F \sim \sqrt{R/k}$, which describes the clearance of the propagation link (see [8]);
(c) the scattering disk scale, $l_3 \sim R/\rho_0 k$, which models the turbulent structure, where R is the length of the radio path.

On the basis on such definitions, Tatarski [19, 45] predicted that the correlation length of the irradiance fluctuations is of the order of the first Fresnel zone $\ell_F \sim \sqrt{R/k}$ (see details in Ref. [8]). However, measurements of the irradiance covariance function under strong fluctuation conditions showed that the correlation length decreases with increasing values of the Rytov variance σ_1^2 and that a large residual correlation tail emerges at large separation distances.

In Refs. [20, 21], the theory developed in Refs. [9, 17–19, 45] was modified for strong fluctuations and showed why the smallest scales of irradiance fluctuations persist into the saturation regime. The basic qualitative arguments presented in these works are still valid. Kolmogorov theory assumes that turbulent eddies range in size from the macroscale to the microscale, forming a continuum of decreasing eddy sizes.

The largest eddy cell size, smaller than that at which turbulent energy is injected into a region, defines an effective outer scale of turbulence L_0, which, near the ground, is roughly comparable with the height of the observation point above ground. An effective inner scale of turbulence l_0 is associated with the smallest cell size before energy is dissipated into heat.

Here, we will briefly present modifications of the Rytov method obtained in Refs. [20, 21] to develop a relatively simple model for irradiance fluctuations that is applicable to moderate to strong fluctuation regimes. In Refs. [20, 21], the following basic observations and assumptions have been stated:

- atmospheric turbulence as it pertains to a propagating wave is statistically inhomogeneous;
- the received irradiance of a wave can be modeled as a modulation process in which small-scale (diffractive) fluctuations are multiplicatively modulated by large-scale (refractive) fluctuations;
- small-scale processes and large-scale processes are statistically independent;
- the Rytov method for signal intensity scintillation is valid even in the saturation regime with the introduction of a spatial frequency filter to account properly for the loss of spatial coherence of the wave in strong fluctuation regimes;
- the geometrical optics method can be applied to large-scale irradiance fluctuations.

These observations and assumptions are based on recognizing that the distribution of refractive power among the turbulent eddy cells of a random medium is described by an inverse power of the physical size of the cell. Thus, the large turbulent cells act as *refractive lenses* with focal lengths typically on the order of hundreds of meters or more, creating the so-called *focusing effect* or *refractive scattering*.

This kind of scattering is defined by the coherent component of the total signal passing the troposphere. The smallest cells have the weakest refractive power and the largest cells the strongest. As a coherent wave begins to propagate into a random atmosphere, the wave is scattered by the smallest of the turbulent cells (on the order of millimeters) creating the so-called *defocusing effect* or *diffractive scattering*. This kind of scattering is defined by the incoherent component of the total signal. Thus, they act as defocusing lenses, decreasing the amplitude of the wave by a significant amount, even for short propagation distances.

As was shown in [5, 8], in the strong fluctuation regime, the spatial coherence radius ρ_0 of the wave determines the correlation length of irradiance fluctuations, and the scattering disk characterizes the width of the residual tail: $R/\rho_0 k$.

The diffractive scattering spreads the wave as it propagates. Refractive and diffractive scattering processes are compound mechanisms, and the total scattering process acts like a modulation of small-scale fluctuations by large-scale fluctuations. Schematically such a situation is sketched in Figure 11.2 containing both components of the total field. Small-scale contributions to scintillation are associated with turbulent cells smaller than the Fresnel zone $\sqrt{R/k}$ or the coherence radius ρ_0, whichever is smaller.

Large-scale fluctuations in the irradiance are generated by turbulent cells larger than that of the first Fresnel zone or the scattering disk $x/k\rho_0$, whichever is larger, and can be described by the method of geometrical optics. Under

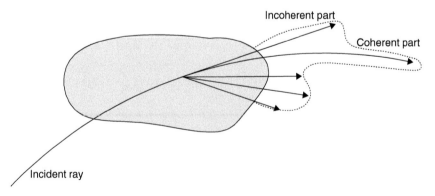

Figure 11.2 Field pattern consisting of the coherent part (I_{co}) and the incoherent part (I_{inc}).

strong fluctuation conditions, spatial cells having size between those of the coherence radius and the scattering disk contribute little to scintillation.

Hence, because of the loss of spatial coherence, only the largest cells nearer the transmitter have focusing effect on the illumination of small diffractive cells near the receiver. Eventually, even these large cells cannot focus or defocus. When this loss of coherence happens, the illumination of the small cells is (statistically) evenly distributed and the fluctuations of the propagating wave are due to random interference of a large number of diffraction scattering of the small eddy cells.

11.4.5 Effects of Atmosphere Turbulences on Signal Fading

The fast fading of the signal at open paths is caused mainly by multipath propagation and turbulent fluctuations of the refractive index. Some very interesting ideas were proposed in [5, 49, 50], which are presented briefly below. As it is known, the fluctuations of the signal intensity due to turbulence are distributed lognormally. For the Kolmogorov model, the normalized standard deviation of this distribution can be presented in terms of C_ε^2 m according to [6, 51, 52], instead of that presented in terms of C_n^2, according to the Rytov's formula (11.31):

$$\sigma_I^2 = 0.12 C_\varepsilon^2 k^{7/6} d^{11/6} \tag{11.32}$$

where $k = 2\pi/\lambda$ is the wave number, $d = R$ is the distance, and C_ε^2 is the structure constant of the turbulence averaged over the path (sign ε sometimes instead of n, because the dielectric permittivity and the refractive index are related as $n^2 = \varepsilon_r$; see Chapter 2). In the atmosphere, the structure constant C_ε^2 may vary within at least four orders of magnitude, from 10^{-15} to 10^{-10} m$^{-2/3}$.

As the path-averaged statistics of these variations is unknown, the margin related to this kind of fading may be estimated only heuristically. The normalized temporal correlation function was obtained in [6, 51, 52]:

$$K(\tau) = \frac{1}{\sin(\pi/12)} \left[(1 + \alpha^4/4)^{11/12} \sin\left(\frac{\pi}{12} + \frac{11}{6} \arctan \frac{\alpha^2}{2} \right) - \frac{11}{6} \left(\frac{\alpha}{\sqrt{2}} \right)^{5/3} \right] \tag{11.33}$$

where $\alpha = \tau/\tau_0$, $\tau_0 = \frac{\sqrt{d/k}}{v}$, and v is the projection of the vehicle velocity to the plane that is perpendicular to the path. The correlation time τ_c defined as $K(\tau_c) = 0.5$ can be estimated as $\tau_c \approx 0.62\tau_0$. The spectrum of the intensity fluctuations according to [49, 50]

$$S(\omega) = \beta^2 w(\omega)/\omega \tag{11.34}$$

was then calculated by using the notion of the normalized spectral density:

$$w(\omega) = 4\omega \int_0^\infty d\tau \; \cos(\omega\tau) K(\tau)$$

which at high and low frequencies is given, respectively, by [6, 51, 52]

$$w(\Omega) = 12.0 \; \Omega^{-5/3}, \quad \Omega \geq 5 \tag{11.35a}$$

and at low frequencies can be approximated as [6, 44, 52]

$$w(\Omega) = 3.47 \; \Omega \; \exp[-0.44 \, \Omega^{\phi(\Omega)}], \quad \Omega \leq 5 \tag{11.35b}$$

where

$$\phi(\Omega) = 1.47 - 0.054 \, \Omega \tag{11.36}$$

and

$$\Omega = \tau_0 \omega \tag{11.37}$$

is the dimensionless frequency. The normalized density $w(\Omega)$ has the maximal value of about 2.30 at $\Omega_m \approx 1.60$, and therefore $\omega_m \approx 1.60/\tau_0$ [50, 53–55]. The phase fluctuations have normal distribution with dispersion

$$\sigma_s^2 = 0.075 \; C_\varepsilon^2 k^2 d \, s_0^{-5/3} \tag{11.38}$$

where $s_0 \sim 2\pi/L_0$, and L_0 is the outer scale of the turbulent spectrum depending on the height and is equal to approximately 10–100 m. Estimations carried out according to Refs. [53, 54] showed that the phase fluctuations caused by turbulence are negligible under typical atmospheric conditions and even for extremely strong turbulence.

11.5 Optical Waves Propagation Caused by Atmospheric Scattering

For radio paths through the atmosphere the dominant propagation mechanism is the scattering from atmospheric turbulent inhomogeneities and discontinuities in the refractive index of the atmosphere. For troposcattering propagation the received signals are generally 50–100 dB below free space values and are characterized by short-term fluctuations superimposed on long-term variations. The statistical distributions are Rayleigh for the short-term and lognormal for the long-term variations (see Chapter 5). The average signal intensity of the scattered signal at the receiving antenna is also given following [49, 50]:

$$I_s = \frac{\pi}{2} k^4 \Phi_\varepsilon(\mathbf{K}_0, 0) \; V_e / r_i^2 r_s^2 \tag{11.39}$$

where $k = 2\pi/\lambda$ is the wave number, r_i is the distance from the optical transmitter to the scattering volume, r_s is the distance from the scattering volume to the optical receiver, V_e is the effective volume of scattering, $\Phi_\varepsilon(\mathbf{K}_0, \mathbf{0})$ is the spectrum of locally homogeneous turbulent permittivity fluctuations at the center of the turbulent zone, and $\Phi_\varepsilon(\mathbf{K}, \mathbf{R})$ is the spectrum of locally homogeneous turbulent permittivity fluctuations at distance $\mathbf{R}$ from its center. The permittivity fluctuations are characterized by the correlation function

$$B_\varepsilon(\mathbf{P}, \mathbf{R}) = \langle \tilde{\varepsilon}(\mathbf{R}_1)\tilde{\varepsilon}(\mathbf{R}_2) \rangle \tag{11.40}$$

where

$$\mathbf{P} = \mathbf{R}_1 - \mathbf{R}_2, \quad \mathbf{R} = \frac{1}{2}(\mathbf{R}_1 + \mathbf{R}_2)$$

and the angular brackets mean ensemble averaging. The spectrum $\Phi_\varepsilon(\mathbf{K}, \mathbf{R})$ is given by the Fourier transform as

$$\Phi_\varepsilon(\mathbf{K}, \mathbf{R}) = (2\pi)^{-3} \int d\mathbf{P} \, \exp(-i\mathbf{K} \cdot \mathbf{P}) B_\varepsilon(\mathbf{P}, \mathbf{R}) \tag{11.41}$$

The expression for the effective scattering volume has the form [3, 49, 50]

$$V_e = \int d\mathbf{R} F(\mathbf{n}_i, \mathbf{n}_s) \Phi_\varepsilon(\mathbf{K}, \mathbf{R}) / \Phi_\varepsilon(\mathbf{K}_0, \mathbf{0}) \tag{11.42}$$

where

$$F(\mathbf{n}_i, \mathbf{n}_s) = |f_i(\mathbf{n}_i)f_s(\mathbf{n}_s)|^2 \tag{11.43}$$

$f_i(\mathbf{n}_i)$ and $f_s(\mathbf{n}_s)$ are the radiation patterns of the transmitting and receiving antennas, respectively. In (11.39) and (11.42), the spatial frequency vector $\mathbf{K}_0$ is defined as

$$\mathbf{K}_0 = k(\mathbf{n}_{i0} - \mathbf{n}_{s0}) \tag{11.44}$$

where the unit vectors $\mathbf{n}_{i0} = \mathbf{r}_{i0}/r_{i0}$ and $\mathbf{n}_{s0} = \mathbf{r}_{s0}/r_{s0}$ are related to the lines connecting the transmitting and receiving terminals with the center of the scattering volume.

By using (11.39) for the intensity of the scattered wave, we can calculate the power received by optical detector as

$$P_2 = F_s^2 G_2 P_1 \tag{11.45}$$

where the scattering loss is given by [3, 49, 50]

$$F_s^2 = \frac{\pi^2}{2} k^2 \Phi_\varepsilon(\mathbf{K}_0, \mathbf{0}) \, V_e / r_i^2 r_s^2 \tag{11.46}$$

There are two unknowns in (11.46). First is the spectrum $\Phi_\varepsilon(\mathbf{K}_0, \mathbf{0})$, which is proportional to the structure parameter of the turbulence, which is characterized by a significant variability (see Section 11.4). The anisotropic structure of the permittivity fluctuations can also cause rather strong variations of the received power.

Second, the effective scattering volume depends essentially on the radiation patterns of both the receiver and the transmitter. Moreover, for sources with relatively small amplification, such as those located at the air vehicle, Eq. (11.42) for the effective scattering volume is no longer valid and should be corrected [51, 52, 56].

What is very important is that it is the frequency selectivity of the channels formed by the tropospheric scattering from turbulences (see Section 11.4). Therefore, to complete the evaluation of the link budget and frequency selectivity for the atmospheric optical paths, the realistic models of the atmospheric turbulence, including anisotropic layered structures, as well as the real radiation patterns of the laser beam should be taken into account.

Concluding the material described in this chapter, we can summarize that for all effects of hydrometeors (rain, clouds, and fog), as well as fast fading caused by atmospheric turbulences, multipath phenomena due to atmospheric inhomogeneities and diffuse scattering should be taken into account in land–atmospheric, or atmospheric–atmospheric optical communication links. The effects of the atmospheric link and its "response" on signal data stream propagation via such kinds of wireless channels with frequency-selective slow and fast fading will be discussed in Chapter 12.

References

1 Pruppacher, H.R. and Pitter, R.L. (1971). A semi-empirical determination of the shape of cloud and rain drops. *J. Atmos. Sci.* 28: 86–94.

2 Slingo, A. (1989). A GSM parametrization for the shortwave radiative properties of water clouds. *J. Atmos. Sci.* 46: 1419–1427.

3 Chou, M.D. (1998). Parametrizations for cloud overlapping and shortwave single scattering properties for use in general circulation and cloud ensemble models. *J. Clim.* 11: 202–214.

4 International Telecommunication Union, ITU-R Recommendation P. 840-2 (1997). *Attenuation due to clouds and fog*. Geneva.

5 Liou, K.N. (1992). *Radiation and Cloud Processes in the Atmosphere*. Oxford, England: Oxford University Press.

6 Blaunstein, N., Arnon, S., Zilberman, A., and Kopeika, N. (2010). *Applied Aspects of Optical Communication and LIDAR*. New York: CRC Press, Taylor & Francis Group.

7 Bean, B.R. and Dutton, E.J. (1966). *Radio Meteorology*. New York: Dover.

8 Blaunstein, N. and Christodoulou, C. (2007). *Radio Propagation and Adaptive Antennas for Wireless Communication Links: Terrestrial, Atmospheric and Ionospheric*. New Jersey: Wiley InterScience.

9 International Telecommunication Union, ITU-R Recommendation, P. 838 1992. *Specific attenuation model for rain for use in prediction methods.* Geneva.

10 Saunders, S.R. (1999). *Antennas and Propagation for Wireless Communication Systems.* New York: Wiley.

11 Twomey, S. (1977). *Atmospheric Aerosols.* Amsterdam: Elsevier.

12 McCartney, E.J. (1976). *Optics of the Atmosphere: Scattering by Molecules and Particles.* New York: Wiley.

13 Whitby, K.Y. (1978). The physical characteristics of sulfur aerosols. *Atmos. Environ.* 12: 135–159.

14 Friedlander, S.K. (1977). *Smoke, Dust and Haze.* New York: Wiley.

15 Seinfeld, J.H. (1986). *Atmospheric Chemistry and Physics of Air Pollution.* New York: Wiley.

16 d'Almeida, G.A., Koepke, P., and Shettle, E.P. (1991). *Atmospheric Aerosols, Global Climatology and Radiative Characteristics.* Hampton: Deepak Publishing.

17 Ishimaru, A. (1978). *Wave Propagation and Scattering in Random Media.* New York: Academic Press.

18 Rytov, S.M., Kravtsov, Y.A., and Tatarskii, V.I. (1988). *Principles of Statistical Radiophysics.* Berlin: Springer.

19 Tatarski, V.I. (1961). *Wave Propagation in a Turbulent Medium.* New York: McGraw-Hill.

20 Andrews, L.C. and Phillips, R.L. (2005). *Laser Beam Propagation through Random Media*, 2e. Bellingham, WA, USA: SPIE Press.

21 Kopeika, N.S. (1998). *A System Engineering Approach to Imaging.* Bellingham, WA: SPIE Press.

22 US Standard Atmosphere (1976). US GPO, Washington, D.C.

23 Kovalev, V.A. and Eichinger, W.E. (2004). *Elastic Lidar: Theory, Practice, and Analysis Methods.* Hoboken, New Jersey: Wiley.

24 Junge, C.E. (1963). *Air Chemistry and Radioactivity.* New York: Academic Press.

25 Jaenicke, R. (1988). Aerosol physics and chemistry. In: *Physical Chemical Properties of the Air, Geophysics and Space Research*, vol. 4 (b) (ed. G. Fisher). Berlin: Springer-Verlag.

26 d'Almeida, G.A. (1987). On the variability of desert aerosol radiative characteristics. *J. Geophys. Res.* 93: 3017–3026.

27 Shettle, E.P. (1984). Optical and radiative properties of a desert aerosol model. In: *Proceedings of the Symposium on Radiation in the Atmosphere* (ed. G. Fiocco), 74–77. A. Deepak Publishing.

28 Remer, L.A. and Kaufman, Y.J. (1998). Dynamic aerosol model: Urban/Industrial aerosol. *J. Geophys. Res.* 103: 13859–13871.

29 Crutzen, P.J. and Andreae, M.O. (1990). Biomass burning in the tropics: impact on atmospheric chemistry and biogeochemical cycles. *Science* 250: 1669–1678.

30 Rosen, J.M. and Hofmann, D.J. (1986). Optical modeling of stratospheric aerosols: present status. *Appl. Opt.* 25 (3): 410–419.

31 Butcher, S.S. and Charlson, R.J. (1972). *Introduction to Air Chemistry*. New York: Academic Press.

32 Cadle, R.D. (1966). *Particles in the Atmosphere and Space*. New York: Van Nostrand Reinhold.

33 Shettle, E.P. and Fenn, R.W. 1979. Models for the aerosols of the lower atmosphere and the effects of humidity variations on their optical properties, AFGL-TR-79-0214.

34 Herman, B., LaRocca, A.J., and Turner, R.E. (1989). Atmospheric scattering. In: *Infrared Handbook* (eds. W.L. Wolfe and G.J. Zissis). Environmental Research Institute of Michigan.

35 Zuev, V.E. and Krekov, G.M. (1986). *Optical Models of the Atmosphere*. Leningrad: Gidrometeoizdat.

36 Hobbs, P.V., Bowdle, D.A., and Radke, L.F. (1985). Particles in the lower troposphere over the high plains of the United States. I: size distributions, elemental compositions and morphologies. *J. Clim. Appl. Meteorol.* 24: 1344–1349.

37 Kent, G.S., Wang, P.-H., McCormick, M.P., and Skeens, K.M. (1995). Multiyear stratospheric aerosol and gas experiment II measurements of upper tropospheric aerosol characteristics. *J. Geophys. Res.* 98: 20725–20735.

38 Berk, A., Bernstein, L.S., and Robertson, D.C. (1989). MODTRAN: A moderate resolution model for LOWTRAN 7, *Air Force Geophys. Lab. Tech. Rep. GL TR-89-0122*. Hanscom AFB, MA.

39 Jursa, A.S. (ed.) (1985). *Handbook of Geophysics and the Space Environment*. Air Force Geophysics Laboratory.

40 Butcher, S.S. and Charlson, R.J. (1972). *Introduction in Air Chemistry*. New York: Academic Press.

41 McClatchey, R.A., Fenn, R.W., Selby, J.E.A. et al. (1972). *Optical Properties of the Atmosphere*. Bedford, MA, AFCRL – 72 – 0497: Air Force Cambridge Res. Lab.: L.G. Hanscom Field.

42 Deirmenjian, D. (1969). *Electromagnetic Scattering On Spherical Polydispersions*. New York: American Elsevier Publishing.

43 Richardson, L.F. (1922). *Weather Prediction by Numerical Process*. Cambridge, UK: Cambridge University Press.

44 Kolmogorov, A.N. (1961). The local structure of turbulence in incompressible viscous fluids for very large Reynolds numbers. In: *Turbulence, Classic Papers on Statistical Theory* (eds. S.K. Friedlander and L. Topper), 151–155. New York: Wiley-Interscience.

45 Tatarskii, V.I. (1971). *The Effects of the Turbulent Atmosphere on Wave Propagation.* Jerusalem: Trans. For NOAA by the Israel Program for Scientific Translations.

46 Kraichman, R.H. (1974). On Kolmogorov's inertial-range theories. *J. Fluid Mech.* 62: 305–330.

47 Obukhov, A.M. (1949). Temperature field structure in a turbulent flow. *Izv. Acad. Nauk SSSR Ser. Geog. Geofiz.* 13: 58–69.

48 Doluhanov, M.P. (1972). Propagation of radio waves. *Moscow: Science.*

49 Samelsohn, G.M. (1993). Effect of inhomogeneities' evolution on time correlation and power spectrum of intensity fluctuations of the wave propagating in a turbulent medium. *Sov. J. Commun. Technol. Electron.* 38: 207–212.

50 Samelsohn, G.M. and Frezinskii, B.Y. (1992). *Propagation of Millimeter and Optical Waves in a Turbulent Atmosphere.* St.-Petersburg: Telecommunication University Press.

51 Bendersky, S., Kopeika, N., and Blaunstein, N. (2004). Atmospheric optical turbulence over land in middle-east coastal environments: prediction, modeling and measurements. *J. Appl. Opt.* 43: 4070–4079.

52 Blaunstein, N. and Kopeika, N. (eds.) (2018). *Optical Waves and Laser Beams in the Irregular Atmosphere.* Boca Raton, FL: CRC Press, Taylor and Frances Group.

53 Bello, P.A. (1969). A troposcatter channel model. *IEEE Trans. Commun.* 17: 130–137.

54 Stremler, F.G. (1982). *Introduction to Communication Systems.* Reading: Addison-Wesley.

55 Champagne, F.H., Friehe, C.A., LaRye, J.C., and Wyngaard, J.C. (1977). Flux measurements, flux-estimation techniques, and fine-scale turbulence measurements in the unstable surface layer over land. *J. Atmos. Sci.* 34: 515–530.

56 Bendersky, S., Kopeika, N., and Blaunstein, N. (2004). Prediction and modeling of line-of-sight bending near ground level for long atmospheric paths. In: *Proceedings of the SPIE International Conference*, 512–522. San Diego, California, San Dieg (August 3–8): Optical Society of America.

Part V

Data Stream Parameters in Atmospheric and Fiber Optic
Communication Links with Fading

12

Transmission of Information Data in Optical Channels: Atmospheric and Fiber Optics

As was shown in Chapter 11, in wireless atmospheric optical communication links, the media vary randomly in the time and space domains. In such situations, the amplitude and phase of optical signals with data, passing such channels, can fluctuate randomly and similarly in the time and space domains, according to the ergodicity of the process of propagation via multipath channel with fading [1–3]. Recently, considerable efforts were devoted to study the influences of atmospheric effects on wave propagation, optical or/and radio, propagating along or through the atmosphere [3–13]. Many other cases of degrading propagation phenomena occur within atmospheric wireless communication links, such as attenuation, optical signal amplitude, and phase scintillations caused by atmospheric turbulence [14–22], absorption by gases, molecular absorption, scattering and attenuation by hydrometeors (fog, smoke, rain, snow, and so on), and scattering by irregularities of refractive index [10, 21, 23] (see also Chapter 11). As for noises in atmospheric communication links and effects on the data stream parameters passing through such channels, their main source is turbulence, which causes signal scintillation or *fading* [11, 13, 23–27]. Among these atmospheric phenomena, atmospheric turbulence plays a significant role in the creation of the distortion to the received signal [14–22].

In order to achieve high performance along with high data rate and increase in channel capacity parameters including minimization of bit error rate (*BER*), it is important to decrease the impact of atmospheric processes, which influence optical signal propagation within the wireless communication links. The definition of the capacity of any communication links is given in the classical communication theory [1–3] as a *traffic load of data in bits per second*.

Several approaches were proposed during the recent decades where the researchers tried to evaluate the data capacity of the communication link, radio and optical, based on the Shannon theorem [28–34]. In the literature, this approach is usually called *classical* [1–3, 11, 13, 14, 28–34]. This approach is usually used for estimation of the capacity in the channel with additive white Gaussian noise (AWGN), that is, with a noise that regularly and constantly

Fiber Optic and Atmospheric Optical Communication, First Edition.
Nathan Blaunstein, Shlomo Engelberg, Evgenii Krouk, and Mikhail Sergeev.
© 2020 John Wiley & Sons, Inc. Published 2020 by John Wiley & Sons, Inc.

distributes along the bandwidth of the communication system or channel [1–3, 11, 13, 14], and is described by Gaussian *PDF* and *CDF* (see all definitions and the corresponding formulas in Chapter 6). There is another approach that takes into account not only AWGN but also multiplicative noise caused by fading phenomena, slow and fast (in time domain) or large-scale and small-scale (in space domain) [1–3, 11, 13, 14] (see also definitions in Chapter 6).

The relation between Gamma-Gamma *PDF*, usually adapted for atmospheric optical communication links [8–10], and Ricean *PDF*, usually used in radio multipath fading communication links [11, 13], using the relation between the signal scintillation parameter σ_I^2 and the Ricean K parameter, was found in Refs. [14, 35, 36], to unify the stochastic approaches for optical and radio communication links (see also Chapter 6). Below, we present the corresponding formulas via Ricean parameter K, which were evaluated for calculation of the data stream parameters based on estimations of the refraction index parameter C_n^2 of turbulences, using experimental data from the corresponding atmospheric measurements described in [13, 14, 35, 36].

12.1 Characteristics of Information Signal Data in Optical Communication Links

According to the classical approach, the *capacity* of the AWGN channel of bandwidth B_ω is based on the Shannon–Hartley formula, which defines the relation between the maximum data rate via any channel, called the capacity, the bandwidth B_ω (in Hz), and the signal-to-noise ratio (SNR $\equiv N_{add}$) (see also Chapter 7):

$$C = B_\omega \log_2 \left[1 + \frac{S}{N_0 B_\omega} \right] \tag{12.1}$$

where in our notations, the power of additive noise in the AWGN channel is $N_{add} = N_0 B_\omega$, S is the signal power, and N_0 is the signal power spectrum (in W/Hz).

According to Refs. [8, 13, 14], signal-to-noise ratio (SNR) equals the ratio between the signal power measured at the input of the receiver, P_R, and its inner white noise N_R, or in decibels:

$$SNR = 10 \log_{10} \left(\frac{P_R}{N_R} \right) = P_{R[dB]} - N_{R[dB]} \tag{12.2}$$

Usually, in radio and optical communication, another characteristic is introduced, called the *spectral efficiency* of the channel/system [1–3, 13, 14, 28–36]:

$$\tilde{C} = \frac{C}{B_\omega} = \log_2 \left[1 + \frac{S}{N_0 B_\omega} \right] \tag{12.3}$$

Based on the second approach (called the *approximate* [35, 36]), account-ing for the fading phenomena, flat or multiselective, we can now estimate the multiplicative noise by introducing a spectral density, N_{mult}, and its frequency bandwidth, B_Ω, in the denominator of the logarithmic function of Eq. (12.2), that is,

$$C = B_\omega \log_2 \left[1 + \frac{S}{N_0 B_\omega + N_{mult} B_\Omega} \right] \tag{12.4}$$

where B_Ω is the frequency bandwidth of the multiplicative noise.

Now, we rewrite formula (12.4) describing the capacity of the channel as

$$C = B_w \log_2 \left(1 + \frac{S}{N_{add} + N_{mult}} \right) = B_w \log_2 \left(1 + \left(\frac{N_{add}}{S} + \frac{N_{mult}}{S} \right)^{-1} \right) \tag{12.5}$$

where $N_{add} = N_0 B_\omega$ and $N_{mult} = N_{mult} B_\Omega$. We can now rewrite Eq. (12.5) by introducing in it Ricean K-factor of fading, defined in Chapter 6, as the ratio of the coherent and multipath (incoherent) components of signal intensity, that is, $K = I_{co}/I_{inc}$, or following special definitions made in [13, 14, 35, 36], $K = S/N_{mult}$. Using these notations, we finally get the capacity as a function of the K-factor and the signal to additive noise ratio (SNR$_{add}$):

$$C = B_\omega \log_2 \left(1 + \left(SNR_{add}^{-1} + K^{-1} \right)^{-1} \right) = B_\omega \log_2 \left(1 + \frac{K \cdot SNR_{add}}{K + SNR_{add}} \right) \tag{12.6}$$

Consequently, it is easy to obtain from Eq. (12.3) the spectral efficiency of the channel:

$$\tilde{C} = \frac{C}{B_\omega} = \log_2 \left(1 + \frac{K \cdot SNR_{add}}{K + SNR_{add}} \right) \tag{12.7}$$

where the bandwidth B_ω changes according to the system under investigation. A comparison made in [35, 36] between the two approaches, classical and approximate, showed that formulas (12.1) with $N_{add} = N_0 B_\omega$ and (12.6) give the same description of the channel/system capacity, when the K-factor is larger than SNR$_{add}$.

Let us show this following Ref. [36]. We compare two approaches, the classi-cal and the approximate, by analyzing changes in the capacity as a function of K, i.e. as a function of different conditions of optical wireless channel, using Eqs. (12.1) (without K-factor) and (12.6) (using K-factor) by examining the spec-tral efficiency (12.7) for three different typical values of SNR$_{add}$. The results of the computations are shown in Figure 12.1. Here, the dashed curves describe results of computations according to the classical approach described by (12.1) normalized on B_ω according to the definition of spectral efficiency introduced

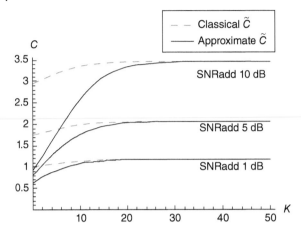

Figure 12.1 Spectral efficiency as a function of K-factor.

above. Continuous curves describe the results of computations based on the approximate approach using formula (12.7). All curves are depicted for different values of signal to AWGN ratio denoted as above by SNR_{add}. It is clearly seen that with the increase in SNR_{add} (from 1 to 10 dB) the spectral efficiency is increased by more than three times. It is also clear that both approaches are close when $K \gg 1$. This effect is more vivid for the additive noise, which is less than parameter K in decibels, i.e. for $SNR_{add} \leq 10$ dB.

We can now express K-factor vs. the capacity or the spectral efficiency, following results obtained in Refs. [14, 35–37], using relations (12.6) and (12.7), respectively. Then, we get

$$K = \frac{SNR_{add}(2^{C/B_w} - 1)}{SNR_{add} - (2^{C/B_w} - 1)} = \frac{SNR_{add}(2^{\tilde{C}} - 1)}{SNR_{add} - (2^{\tilde{C}} - 1)} \tag{12.8}$$

Equation (12.8) is an important result, since it gives a relation between the spectral efficiency of the multipath optical communication channel, wired (e.g. fiber) or wireless, caused by fading phenomena, and the K-factor of fading occurring in such a channel.

Finally, we can relate the strength of the scintillation, introduced in Refs. [14, 35–37], which is characterized by the normalized intensity variance, $\langle \sigma_I^2 \rangle$, called the *scintillation index*, with the K-factor of fading. In Chapter 11, we presented this characteristic, following Refs. [13, 14, 35–37], for zero-mean random process. For the reader's convenience to understand the subject, we present this relation again:

$$\langle \sigma_I^2 \rangle = \frac{\langle [I - \langle I \rangle]^2 \rangle}{\langle I \rangle^2} = \frac{I_{inc}^2}{I_{co}^2} \equiv K^{-2} \tag{12.9}$$

where I_{co} and I_{inc} are the coherent and incoherent components of the total signal intensity.

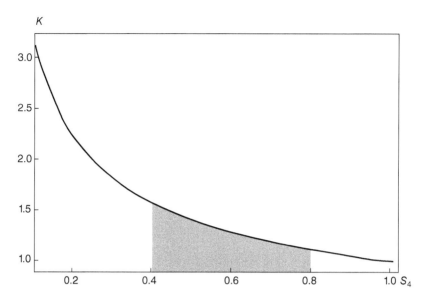

Figure 12.2 K-factor of fading vs. the intensity scintillation index (according to [37]). Source: Reproduced with permission of Mikhail O. Samolovov from SUAI Publishing Center.

We can rearrange Eq. (12.9) in such a manner to obtain relations between K-factor and σ_I^2, denoted as in Refs. [36–38] by S_4, which is more convenient, assuming that the latter can be obtained via its relations with C_n^2, which was estimated experimentally in [14] for any scenario occurring in the turbulent nonhomogeneous atmosphere:

$$K = (\sigma_I^2)^{-1/2} \equiv S_4^{-1/2} \tag{12.10}$$

Dependence of K-factor of fading vs. the signal intensity scintillation σ_I^2, usually denoted by S_4, is presented in Figure 12.2, according to [37]. The range $\langle \sigma_I^2 \rangle$ of the scintillation index variations, from 0.4 to 0.8, is obtained from numerous experiments where relations between this parameter and the refractivity of the turbulence in the nonhomogeneous atmosphere are taken into account (see Refs. [13, 14, 35–38]). Thus, from experiments described there, it is estimated to be around $C_n^2 \approx 10^{-15}$ m$^{-2/3}$ and around $C_n^2 \approx 10^{-13}$ m$^{-2/3}$ for nocturnal and daily periods, respectively.

It can be seen from Figure 12.2 that when the mean and strong atmospheric turbulences occur in the nonhomogeneous atmosphere, where the index of signal intensity scintillations varies from 0.4 to 0.9, the corresponding K-parameter varies, according to Eq. (12.10), in the range of about 1.1–1.6.

It indicates the existence of direct visibility (i.e. the LOS component) between both terminals, the source and the detector, accompanied by the additional effects of multipath phenomena (i.e. NLOS multipath component) caused by multiple scattering of optical rays at the turbulent structures formed in the perturbed atmospheric regions, observed experimentally (see Refs. [35–37]). In such scenarios, where $S_4 \equiv \sigma_I^2 = 0.7 - 0.9$, the K-factor, described as multipath fading phenomenon within the atmospheric wireless propagation channel, changes around the unit. Having now information about the K-factor, we can predict deviations of the parameters of the data stream (i.e. the signal) in the multipath channels passing through the strong turbulences occurring in the nonhomogeneous atmosphere.

Thus, the capacity or spectral efficiency described vs. K-factor by Eqs. (12.6) and (12.7), respectively, can be easily estimated for various scenarios occurring in the atmospheric channel and for different conditions of the inner noise of the optical transmitters and receivers inside the optical communication under consideration. One of the examples can be seen in Figure 12.3, according to Ref. [37] for different additive SNRs and for a "point" optical beam (with respect to the diameter of the detector).

We take this parameter much more wider, varying in the interval from 0.1 to 700, that is, to cover the "worst" case, when $K \ll 1$, which is described by

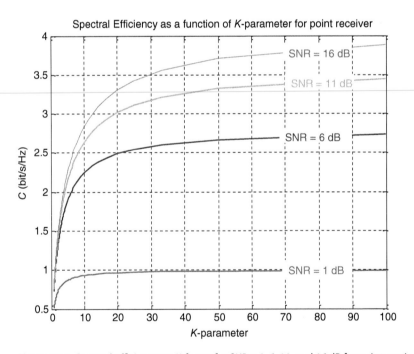

Figure 12.3 Spectral efficiency vs. K-factor for $SNR = 1, 6, 11,$ and 16 dB for point receiver. Range between the terminals is $L = 1$ km (extracted from [37]). Source: Reproduced with permission of Mikhail O. Samolovov from SUAI Publishing Center.

Rayleigh law (see Chapter 6) passing the quasi-LOS case, when $K \approx 1$, and finally achieving a situation where $K \gg 1$ (ideal LOS case in propagation, described by Delta-shape Gaussian law). As can be seen, in the strong perturbed nonhomogeneous atmosphere (with strong turbulences), where $K \leq 1$, the spectral efficiency is around 0.7–0.8 (for SNR = 1 dB) and 1.6–2.0 (for other values of SNR = 6, 11, 12 dB).

12.2 Bit Error Rate in Optical Communication Channel

Usually, wireless or fiber optic communication systems use well-known modulation techniques for encoding the received signal [1–3]. While applying the coherent detection process for encoding (see Chapters 4 and 5) of information into amplitude, frequency, or phase of the transmitted signal, the usually known modulation techniques are used. They are amplitude shift keying (ASK), frequency shift keying (FSK), and phase shift keying (PSK), when modulation follows changes of amplitude, frequency, and phase, respectively (see Chapter 7).

Below, we will describe the case of ASK modulation, which deals with the signal strength or power as a function of time, as a common method of signal processing usually used in optical detectors (see Chapter 7). Since the optical signal can be considered as a carrier of digital information, as a set of bits, effects of multipath fading in the optical communication channel lead to errors in bits characterized by the special parameter defined as BER. Thus, using the Rayleigh distribution, we can determine the probability of bit error occurring in the multipath channel operating with ASK modulation by the following formula [1–3, 11, 14]:

$$P_r(e) = \frac{1}{\sigma^2} \int_{r_T}^{\infty} r\, e^{-\left[\frac{r^2}{2\sigma_N^2}\right]} dr = e^{-\left[\frac{r_T^2}{2\sigma_N^2}\right]} \tag{12.11}$$

where $P_r(e)$ represents the evaluated probability of a bit error, σ_N^2 is the intensity of interference at the optical receiver (usually determined as the multiplicative noise [35, 36]), and r_T determines the threshold between detection without multiplicative noise (defined as a "good case" [11, 13] and with multiplicative noise (defined as a "bad case" [11, 13]).

The expected probability of error in optical ASK modulation systems was suggested in Refs. [8, 10] as a function of PDF of random variable s, $p_i(s)$, which represents the well-known *Gamma-Gamma* distribution, briefly described in Chapter 6:

$$P_r(e) = \frac{1}{2} \int_0^{\infty} p_i(s) \cdot \text{erfc}\left(\frac{\langle \text{SNR} \rangle \cdot s}{2\sqrt{2}\langle I_s \rangle}\right) ds \tag{12.12}$$

Here, as above, $P_r(e)$ is the evaluated probability of the error, $\text{erfc}(\bullet)$ is the error function probability, both introduced in Chapter 6, and $\langle I_s \rangle$ is the average intensity of the received signal.

In our investigations, we use the well-known relations between σ_I^2, σ_x^2, and σ_y^2, introduced in Chapter 6 according to Gamma-Gamma distribution, with parameters α_0^2 and β_0^2 defined in Refs. [14, 36]. The main goal of this procedure is to show the reader the practical relation between the signal parameter σ_I^2 and the channel parameter, C_n^2, following Refs. [8, 10, 14, 36, 37].

Using the definitions above, according to [8, 10] we can rewrite Eq. (12.12) in the following form:

$$
\text{BER} \equiv P_r(e) = \frac{1}{2} \int_0^\infty \text{erfc} \left(\frac{\text{SNR} \cdot s}{2\sqrt{2\langle I_s \rangle}} \right) \frac{2(\alpha\beta)^{(\alpha+\beta)/2}}{\Gamma(\alpha)\Gamma(\beta)\langle I_s \rangle} \left(\frac{s}{\langle I_s \rangle} \right)^{\frac{(\alpha+\beta)}{2}-1}
$$

$$
K_{(\alpha-\beta)} \left(2\sqrt{\frac{\alpha \cdot \beta \cdot s}{\langle I_s \rangle}} \right) ds \tag{12.13}
$$

where, as above, $\langle I_s \rangle$ represents the average intensity of the received signal, s is the randomly varied statistically independent integration parameter, $K_{(\alpha-\beta)}$ is the modified Bessel function of order $(\alpha - \beta)$, and $\Gamma(\alpha)$ and $\Gamma(\beta)$ are the Gamma functions introduced in Chapter 6 (see also Refs. [8, 10]).

We can also present BER as a function of the Ricean K-factor of fading, following results obtained in Ref. [35]. Thus, we follow Ref. [35] and use a classical formula for BER, according to Refs. [28–34] that yields

$$
\text{BER} = \frac{1}{2} \int_0^\infty p(x) \text{erfc} \left(\frac{\text{SNR}}{2\sqrt{2}} x \right) dx \tag{12.14}
$$

where $p(x)$ is the probability density function, which in our case was taken as Ricean one, and $\text{erfc}(\bullet)$ is the well-known error function (see definitions in Chapter 6). Using the BER definition (12.14), where $p(x)$ is Ricean PDF, and where the SNR includes also the multiplicative noise, we finally get for a *BER* [35]

$$
\text{BER} \left(K, \frac{S}{N_{\text{add}}}, \sigma \right) = \frac{1}{2} \int_0^\infty \frac{x}{\sigma^2} \cdot e^{-\frac{x^2}{2\sigma^2}} \cdot e^{-K} \cdot I_0 \left(\frac{x}{\sigma} \sqrt{2K} \right)
$$

$$
\cdot \text{erfc} \left(\frac{K \cdot \frac{S}{N_{\text{add}}}}{2\sqrt{2} \left(K + \frac{S}{N_{\text{add}}} \right)} x \right) dx \tag{12.15}
$$

This is an important formula that gives the relation between the BER and the additive SNR, the Ricean parameter K described the multipath fading phenomena occurring within the multipath land–satellite communication link passing the atmosphere, and the probability of BER of the information data stream inside such a channel.

12.3 Relations Between Signal Data Parameters and Fading Parameters in Atmospheric Links

We can now enter into a detailed analysis of the key parameters of the optical atmospheric channel and the information data, based on the approximate approach proposed above according to Refs. [14, 35–37].

Graphical representation of the evaluation results was based upon an approach developed on the basis of Eqs. (12.12)–(12.15). The flow diagram introduced in Ref. [37] and shown in Figure 12.4 can easily describe the corresponding algorithm of the proposed approach. Let us explain all steps of this algorithm.

First, the refractive index structure parameter C_n^2, which characterizes the strength of atmospheric turbulence, was measured experimentally. The corresponding procedure describes real experiments [13, 14, 35–37].

In addition, parameter K, which represents the ratio between the coherent and incoherent components of the optical signal within the communication link, can be estimated from the corresponding measurements of fading. Estimations of the intensity of the coherent component were based on the measured transmitted energy divided by the attenuation along the propagation path. At the same time, estimations of the intensity of the incoherent

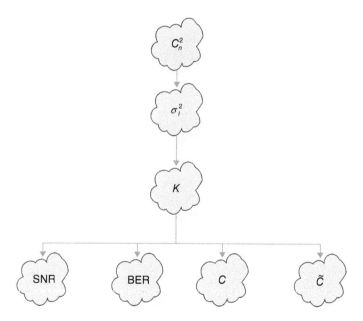

Figure 12.4 The model flow diagram according to [37]. Source: Reproduced with permission of Mikhail O. Samolovov from SUAI Publishing Center.

component were taken from the measured beam scintillation/fluctuation energy, σ_I^2 (see [13, 14, 35–37]). The latter is evaluated based on the relation between σ_I^2, α, β, and α_0^2, β_0^2, as a function of C_n^2 (see [13, 14, 35–37]). Considering these parameters, finally, *SNR* can be calculated. Then, using Eqs. (12.12)–(12.15), the *BER* and the channel spectral efficiency ($\widetilde{C}$) can be evaluated, respectively.

Comparison between *BER* and *SNR* curves over extreme experimental turbulence conditions, $C_n^2 \approx 10^{-15}$ and $C_n^2 \approx 10^{-13}$ m$^{-2/3}$, for daily and nocturnal atmosphere, respectively, for different propagation lengths are shown in Figure 12.5 extracted from Refs. [37]. According to this plot, differences between daily curves for the same atmospheric turbulence value are small in comparison with nocturnal curves gap.

As seen from Figure 12.5, strong turbulence atmospheric conditions, defined by $C_n^2 \approx 10^{-13}$ m$^{-2/3}$, result in relatively great BER and so play a major role in fading effects caused by turbulence [37]. Therefore, there are subjects of particular interest for estimation of optical communication parameters, such as *BER* and *SNR* and prediction of maximal losses caused by strong turbulence.

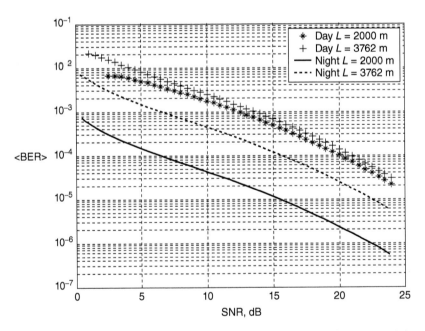

Figure 12.5 Comparison between 2 and 3.76 km propagation lengths for extreme daily and nocturnal experimental turbulence conditions (extracted from [37]). Source: Reproduced with permission of Mikhail O. Samolovov from SUAI Publishing Center.

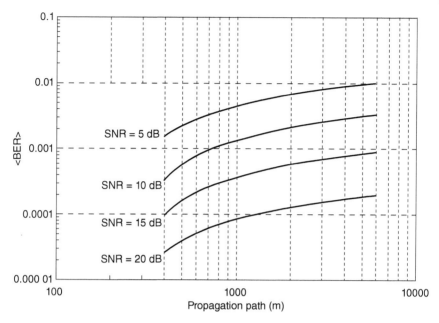

Figure 12.6 *BER* in atmospheric communication links with strong turbulence vs. the optical ray path length, *L*, for different *SNR* values (according to [37]). Source: Reproduced with permission of Mikhail O. Samolovov from SUAI Publishing Center.

Evaluation of *BER* in strong atmospheric turbulence media as a function of propagation length, *L*, for different *SNR* values based on experimental data described in Refs. [14, 35–37] is given by Figure 12.6.

In Figure 12.6, the atmospheric optical communication link ranges are taken from 0.4 to 6 km, which correspond to experiments described in Refs. [14, 35–37]. For less than 0.4 km path the error is negligible, and for over 6 km paths, optical communication systems (lasers) are not usually applied [4, 7–10].

Next, we analyze effects of fading (e.g. the changes of the *K*-parameter) on *BER* conditions within the turbulent wireless communication link consisting of weak and strong turbulences. As shown in Refs. [14, 36, 37], depending on the kind of turbulence – strong (with $\langle C_n^2 \rangle = 5 \cdot 10^{-14} \mathrm{m}^{(-2/3)}$) occurring at altitudes up to 100–200 m, or weak (with $\langle C_n^2 \rangle = 4 \cdot 10^{-16} \mathrm{m}^{(-2/3)}$) occurring at altitudes of 1–2 km – the effects of fading become stronger (at the first case) and weaker (at the second case). In other words, at higher atmospheric altitudes for the horizontal atmospheric channels, where the LOS component exceeds the multipath (NLOS) component (i.e. for $K > 1$), the *BER* characteristic becomes negligible and can be ignored as well as other fading characteristics in the design of land–atmospheric or pure atmospheric links.

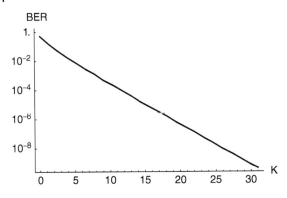

Figure 12.7 BER as a function of K extracted from [37]. Source: Reproduced with permission of Mikhail O. Samolovov from SUAI Publishing Center.

Finally, in Refs. [14, 35–37] an optimal algorithm for minimization of the BER of optical bandpass signals (see definition in Chapter 3) was constructed for different situations occurring in optical atmospheric communication links. Thus, taking some measured data presented in [14, 36], we can show here some examples. In our computations, we used the following parameters of the channel and measured data: $\sigma = 2\,\text{dB}$, and $\text{SNR}_{\text{add}} = 1\,\text{dB}$. The results of the computations are shown in Figure 12.7, extracted from [37] for *BER* as a function of the fading parameter K obtained from the experimental data.

As seen from Figure 12.7 with increase of K parameter, that is, when LOS component becomes predominant with respect to NLOS multipath components, it is found that BER decreases essentially from 10^{-2} for $K \approx 5$ to 10^{-6} for

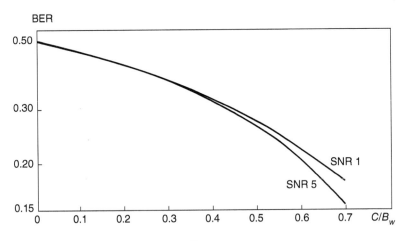

Figure 12.8 BER vs. spectral efficiency $\widetilde{C}$ (according to [37]). Source: Reproduced with permission of Mikhail O. Samolovov from SUAI Publishing Center.

$K \approx 20$ (i.e. for the atmospheric link at altitudes of 100–500 m filled by turbulent structures [37]).

At the same time, as expected in [37], the spectral efficiency was found to increase with the K-parameter. Hence, with the increase of the spectral efficiency of the data stream (from 0.8 to 1.0), a simultaneous sharp decrease of *BER* was also found. To show this, we present in Figure 12.8 *BER* dependence vs. the spectral efficiency.

As can be seen, for small $\tilde{C} = C/B_\omega$, BER is sufficiently high. At the range $\tilde{C} = 0.6 - 0.7$, which is taken from Figure 12.2 as a real range of K-factor in the turbulent atmosphere, the BER is twice smaller compared with the previous case, but it is high enough to lose information inside the channel with fading occurring during the effects of turbulence. It should be noted that as was discussed above and follows from illustrations of Figure 12.3, the increase of SNR inside the channel (from 1 to 5 dB) cannot decrease BER significantly even for high spectral efficiency. Generally speaking, with increase in the spectral efficiency from 0.1 to 0.7, the *BER* parameter decreases approximately three times.

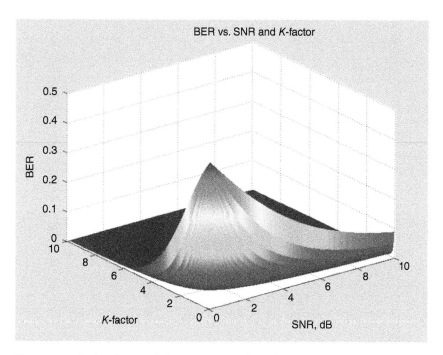

Figure 12.9 3D plot of BER vs. fading parameter K for different SNR varied from 1 to 10 dB.

The above results were generalized both for strong and weak turbulences via measuring of the scintillation index, which allows us to evaluate deviations of the parameter of fading K from $K = 0$ (strong fading) to $K = 10$ (weak fading) for various variations of SNR – from 1 to 10 dB. Results of numerical computations based on experimental data [13, 14, 35–37] are shown in 3D format in Figure 12.9.

From the illustrations shown in Figure 12.9 it is clearly seen that for $0 < K < 1$ and SNR > 5 dB the sharp decrease of BER from $\sim$0.3 to $\sim$0.0003, that is, at three orders, allow us to state a huge decrease in bits' errors inside the data flows passing the optical atmospheric (e.g. wireless) channel.

At the same time, as noted in Ref. [35] and seen from Figure 12.7, the spectral efficiency also increases with increase in fading parameter K. Thus, increase of $\tilde{C} = C/B_\omega$ from 0.2 to 0.8 leads to a sharp decrease of BER from 0.3–0.5 to 0.003–0.005 – an effect that depends on SNR in the optical channel and additionally decreases with increase of SNR.

12.4 Effects of Fading in Fiber Optic Communication Link

As was discussed in Chapter 9, in fiber optic channels fading of optical signals occurs due to two factors: multi-ray phenomena leading to the inter-ray interference (IRI), and due to dispersive properties of the material at the inner and outer coating of the fiber guide caused by the inhomogeneous structure of the wire communication channel. Dispersion of these two types was discussed in Chapter 9, according to the matter presented there, some of the formulas will be presented below for clarity on the subject.

Thus, the time delay dispersion τ of set of pulses along the fiber with the l length, $\Delta(\tau/l)$, can be estimated by knowledge of refraction indexes of the inner and outer cables. n_1 and n_2, that is,

$$\Delta\left(\frac{\tau}{l}\right) = \frac{n_1}{cn_2}(n_1 - n_2) \approx \frac{n_1}{c}\Delta \tag{12.16}$$

In the case of multimode dispersion, a spread of information pulses at the length of optical cable in time is given by (see Chapter 9)

$$\Delta T = \frac{l \cdot n_1^2}{cn_2}\Delta \approx \frac{l \cdot n_1}{cn_2}\Delta \tag{12.17}$$

For the case of dispersion caused by cable material inhomogeneity along the fiber we can use the following formula (see Chapter 9):

$$\Delta\left(\frac{\tau}{l}\right) = -M\Delta\lambda \tag{12.18}$$

All parameters presented in Eqs. (12.17) and (12.18) are fully determined in Chapter 9. We present them here for more evident description of the subject of this chapter.

The limitations in capacity of data flow inside the fiber depend on p–n type of pulses, either return-to-zero (RZ) or non-return-to-zero (NRZ) (see Chapter 4). Thus, for NRZ pulses we get (see also Chapter 9)

$$C_{NRZ} \times l = \frac{0.7}{\Delta(\tau/l)} \tag{12.19}$$

whereas for RZ pulses we get (see Chapter 9)

$$C_{RZ} \times l = \frac{0.35}{\Delta(\tau/l)} \tag{12.20}$$

Using numerical data regarding the material dispersion parameter M presented in Chapter 9, we can also obtain the empirical formulas for computation of the capacity of fiber optic channel at the length l for two types of pulses, that is,

$$C_{NRZ} \times l = 1.75\,[(\text{Mbit/s}) \times \text{km}] \tag{12.21}$$

and

$$C_{RZ} \times l = 0.875\,(\text{Mbit/c}) \times \text{km}] \tag{12.22}$$

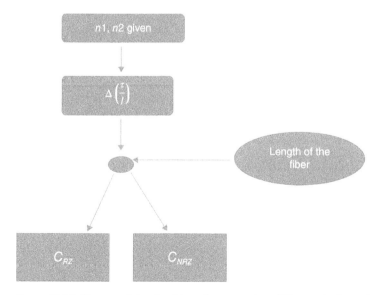

Figure 12.10 Diagram of the algorithm of computation of fiber optic channel capacity in the case of multimode dispersion at the length *l* of the optical cable.

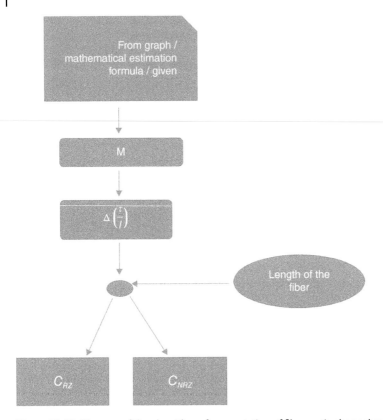

Figure 12.11 Diagram of the algorithm of computation of fiber optic channel capacity in the case of material dispersion M at the length *l* of the optical cable.

In Figures 12.10 and 12.11, the corresponding diagrams shown provide the algorithm of computation of the capacity of the fiber optic communication links with time dispersion caused by the multimode interference (Figure 12.10) and by the material inhomogeneity along the optical cable (Figure 12.11).

Figure 12.12 shows the dependence of the capacity of fiber optic channel vs. the difference of the refraction coefficients of the inner and outer cables at the length of the cable of 1 km and for $n_1 = 1.46$ (according to formula (12.16)).

From Figure 12.12 it is clearly seen that with increase in difference between the refraction indexes of the inner and outer parts of the fiber, the capacity of such a fiber optic channel is decreased exponentially. Thus, the maximum rate of data passing such a channel also decreases exponentially.

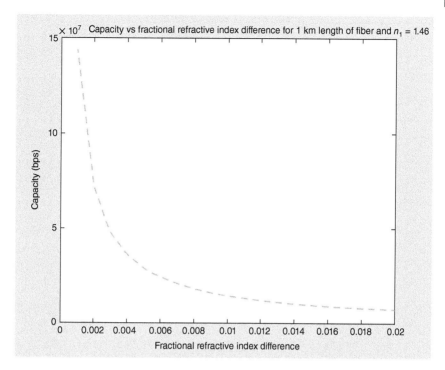

Figure 12.12 The capacity of the fiber communication link with the length of 1 km vs. difference between refraction indexes of the cladding and core of the optical cable for $n_1 = 1.46$.

References

1 Papoulis (1991). *Probability Random Variables and Stochastic Process.* New York: McGraw-Hill.
2 Stuber, G.L. (1996). *Principles of Mobile Communication.* Kluwer.
3 Proakis, J.G. (2001). *Digital Communication,* 4e. McGraw-Hill.
4 Tatarski, V.I. (1961). *Wave Propagation in a Turbulent Medium.* New York: McGraw-Hill.
5 Ishimaru, A. (1978). *Wave Propagation and Scattering in Random Media.* New York: Academic Press.
6 Stremler, F.G. (1982). *Introduction to Communication Systems.* Addison-Wesley Reading.
7 V. A. Banakh and V. L. Mironov, LIDAR in a Turbulence Atmosphere, Artech House, Dedham, 1987.
8 Andrews, L.C. and Phillips, R.L. (1998). *Laser Beam Propagation through Random Media.* Bellingham, WA: SPIE Optical Engineering Press.

9 Kopeika, N.S. (1998). *A System Engineering Approach to Imaging*. Bellingham, WA: SPIE Optical Engineering Press.

10 Andrews, L.C., Phillips, R.L., and Hopen, C.Y. (2001). *Laser Beam Scintillation with Applications*. Bellingham, WA: SPIE Optical Engineering Press.

11 Saunders, S.R. (1999). *Antennas and Propagation for Wireless Communication Systems*. New York: Wiley.

12 Andrews, L.C. (1998). *Special Functions of Mathematics for Engineers*, 2e. Bellingham/Oxford: SPIE Optical Engineering Press/Oxford University Press.

13 Blaunstein, N. and Christodoulou, C. (2007). *Radio Propagation and Adaptive Antennas for Wireless Communication Links: Terrestrial, Atmospheric and Ionospheric*. New Jersey: Wiley InterScience.

14 Blaunstein, N., Arnon, S., Zilberman, A., and Kopeika, N. (2010). *Applied Aspects of Optical Communication and LIDAR*. Boca Raton, FL: CRC Press, Taylor and Frances Group.

15 Kolmogorov, A.N. (1941). The local structure of turbulence incompressible viscous fluid for very large Reynolds numbers. *Rep. Acad. Sci. USSR* 30: 301–305.

16 Monin, A.S. and Obukhov, A.M. (1954). Basic law of turbulent mixing near the ground. *Trans. Akad. Nauk.* 24 (151): 1963–1987.

17 Kopeika, N.S., Kogan, I., Israeli, R., and Dinstein, I. (1990). Prediction of image propagation quality through the atmosphere: the dependence of atmospheric modulation transfer function on weather. *Opt. Eng.* 29 (12): 1427–1438.

18 Sadot, D. and Kopeika, N.S. (1992). Forecasting optical turbulence strength on basis of macroscale meteorology and aerosols: models and validation. *Opt. Eng.* 31: 200–212.

19 Sadot, D., Shemtov, D., and Kopeika, N.S. (1994). Theoretical and experimental investigation of image quality through an inhomogeneous turbulent medium. *Waves Random Media* 4 (2): 177–189.

20 Hutt, D.L. (1999). Modeling and measurements of atmospheric optical turbulence over land. *Opt. Eng.* 38 (8): 1288–1295.

21 Bedersky, S., Kopeika, N., and Blaunstein, N. (2004). Atmospheric optical turbulence over land in middle east coastal environments: prediction modeling and measurements. *J. Appl. Opt.* 43: 4070–4079.

22 Bendersky, S., Kopeika, N., and Blaunstein, N. (2004). Prediction and modeling of line-of-sight bending near ground level for long atmospheric paths. In: *Proceedings of SPIE International Conference*, 512–522. San Diego, California, San Diego, USA (3–8 August): Optical Society of America.

23 Zhang, W., Tervonen, J.K., and Salonen, E.T. (1996). Backward and forward scattering by the melting layer composed of spheroidal hydrometeors at 5–100 GHz. *IEEE Trans. Antennas Propag.* 44: 1208–1219.

24 Macke, A. and Mishchenko, M. (1996). Applicability of regular particle shapes in light scattering calculations for atmospheric ice particles. *Appl. Opt.* 35: 4291–4296.

25 Hovenac, E.A. (1991). Calculation of far-field scattering from nonspherical particles using a geometrical optics approach. *Appl. Opt.* 30: 4739–4746.

26 Spinhirne, J.D. and Nakajima, T. (1994). Glory of clouds in the near infrared. *Appl. Opt.* 33: 4652–4662.

27 Duncan, L.D., Lindberg, J.D., and Loveland, R.B. (1980). An empirical model of the vertical structure of German fogs, ASL-TR-0071, US Army Atmospheric Sciences Laboratory, White Sands Missile Range, N. Mex.

28 Goldsmith, A.J., Greenstein, L.J., and Foschini, G.L. (1994). Error statistics of real-time power measurements in cellular channels with multipath and shadowing. *IEEE Trans. Veh. Technol.* 43 (3): 439–446.

29 Goldsmith, A.J. (1997). The capacity of downlink fading channels with variable rate and power. *IEEE Trans. Veh. Technol.* 46 (3): 569–580.

30 Goldsmith, A.J. and Varaiya, P.P. (1997). Capacity of fading channels with channels side information. *IEEE Trans. Inf. Theory* 43 (6): 1986–1992.

31 Biglieri, E., Proakis, J., and Shamai, S. (1998). Fading channels: information theoretic and communication aspects. *IEEE Trans. Inf. Theory* 44 (6): 2619–2692.

32 Alouini, M.-S., Simon, M.K., and Goldsmith, A.J. (2001). Average BER performance of single and multi carrier DS-CSMA systems over generalized fading channels. *Wiley J. Wirel. Syst. Mob. Comput.* 1 (1): 93–110.

33 Telatar, I.E. and Tse, D.N.C. (2000). Capacity and mutual information of wideband multipath fading channels. *IEEE Trans. Inf. Theory* 46 (4): 1384–1400.

34 Winters, J.H. (1987). On the capacity of radio communication systems with diversity in a Rayleigh fading environments. *IEEE Sel. Areas Commun.* 5: 871–878.

35 Yarkoni, N. and Blaunstein, N. (2006). Capacity and spectral efficiency of MIMO wireless systems in multipath urban environments with fading. In: *Proceedings of European Conference on Antennas and Propagation*, 316–321. San Diego, California, Nice, France (6–10 November): Optical Society of America.

36 Tiker, A., Yarkoni, N., Blaunstein, N. et al. (2007). Prediction of data stream parameters in atmospheric turbulet wireless communication links. *Appl. Opt.* 46 (2): 190–199.

37 Blaunstein, N.S., Kruk, E.A., and Sergeev, M.B. (2016). *Opticheskaya Svyaz' – Optovolokonnaya i Atmosphernaya [Optical Communication – Fiber Optic and Atmospheric]*, 286. Sankt Petersburg: GUAP (in Russian).

38 Blaunstein, N. and Plohotniuk, E. (2008). *Ionosphere and Applied Aspects of Radio Communication and Radar*. New York: CIR Press, Taylor and Frances.

Index

Fiber Optic and Atmospheric Optical Communication, First Edition.
Nathan Blaunstein, Shlomo Engelberg, Evgenii Krouk, and Mikhail Sergeev.
© 2020 John Wiley & Sons, Inc. Published 2020 by John Wiley & Sons, Inc.